TAKE CHARGE OF YOUR HOME RENOVATION

Susan Boyle Hillstrom

AND THE EDITORS OF

House Beautiful

TAKE CHARGE OF YOUR HOME RENOVATION

EVERYTHING YOU NEED TO KNOW FOR A SUCCESSFUL HOME RENOVATION OR REMODELING

HEARST BOOKS

New York

Library of Congress Cataloging-in-Publication Data
Hillstrom, Susan Boyle
Take charge of your home renovation : everything you need to know to manage a successful home renovation or remodeling / Susan Boyle Hillstrom.
p. cm.
ISBN 1-58816-015-7
1. Dwellings—Remodeling. 2. Contractors—Selection and appointment. I. Title.

TH4816 .H548 2001
690'.837—dc21 00-063209

Design: Kathleen Lake, Neuwirth & Associates, Inc.

Printed in The United States of America

First U.S. Edition

1 2 3 4 5 6 7 8 9 10

To subscribe to House Beautiful, please visit us at www.housebeautiful.com or www.hearstmags.com.

Contents

	Acknowledgments	vii
	Introduction	ix
	Before You Begin	xv
one	Who Do You Need to Hire?	1
two	Budgeting and Finding Contractors	29
three	Interviewing the Professionals	63
four	Ironclad Contracts	98
five	The Job Begins	129
six	Troubleshooting	152
seven	Resources	161
	Glossary	186

Acknowledgments

N'ann Harp of Smart Consumer Services was a great help. If she were in charge of the remodeling industry, things would go more smoothly and be a lot more fun.

I thank my husband, Roger Hillstrom, for his intelligent input and for making dinner so often.

Special thanks also to Bruce Commandeur, Jim DeWitt, Laura Glassman, Charlotte Hill, Carol Little, Debbie and Frank Lumia, Carla McClinton, Bruce McKeegan, Ann Rankin, Rick Shaver, Keith Slater, Cynthia Smith, Mike Vella—and to my father, Vilas J. Boyle.

Introduction

I am married to a man who operated a small contracting business for about 20 years, until he moved on to a less physically taxing field of endeavor. Therefore I know that contractors are not by nature unscrupulous people, that many of them, in fact, have high ethical standards, work hard, and often devise wonderfully creative ways to solve the problems that always seem to crop up on a construction site. I was always puzzled—and sometimes embarrassed—by the glee with which some people would launch into horror stories about the trouble they or their friends or their relatives had experienced with contractors, who, in these stories, were always entirely to blame and entirely wicked.

I realized that it is a widely held assumption that contractors—pretty much all contractors—are sleazy. Bring together a bunch of people who have never met before and who have very little in common and that's one of the few things they'll agree upon. Even some of the research I did for this book revealed an anti-contractor bias among sources of information aimed at homeowners. A few web sites I encountered, for example, started off with a premise like this: "Here's the information you need to protect you against the evil contractors that are lurking, lying in wait, hoping to rip you off."

But most of the research and interviews I have conducted showed me a very different picture.

Yes, bad things do happen to good homeowners. And those bad things are much more likely to happen when those good homeowners are also careless homeowners who fail to take the time or trouble to fully and carefully check out prospective contractors, and who after they've found a solid one, fail to insist on an equally solid contract. Advice abounds—in books, magazines, newspapers, on the Internet, even on television—but much of it is superficial and not truly helpful. As a result, many people don't really know how to choose a contractor wisely and how to protect themselves for the duration of the project and beyond. With this book, I hope to change that and to help make your home renovation the exciting, satisfying experience it should be, in spite of the inevitable dust and difficulties.

Every year, billions of dollars are spent on home remodeling, and the figures regularly get larger. The dollars spent on remodeling in 1998 represented about 65 percent of total spending on new home construction. It is expected that the remodeling dollar volume will overtake new housing expenditures in the next decade. In 1999, home-improvement spending reached an astonishing $135 billion, according to the National Association of the Remodeling Industry (NARI) in Alexandria, Virginia. Spending in 2000, says NARI, is expected to exceed $150 billion, about an 8 percent increase. The National Association of Home Builders (NAHB) in Washington, D.C., predicts that by 2006 the remodeling market will increase to about $168 billion. The surge in the remodeling market is a result of several factors—"A healthy economy, a population increase, an aging housing stock, and more people owning their own homes"—says Ken Skowronski, a remodeling contractor and president of NARI.

That's good news for everybody—homeowners, banks, remodeling contractors, lumberyards and other suppliers, and manufacturers of everything from kitchen appliances to roofing shingles to windows to paint.

Waiting for the bad news? Here it is. Each year incalculable millions of dollars are spent righting the wrongs that occur when a remodeling goes bad—repairing botched work, replacing supplies, hiring a second contractor to finish the job, and, sadly, hiring a lawyer and going to court. In the dizzying dance of remodeling and construction litigation, homeowners sue contractors, subcontractors, or suppliers; contractors sue homeowners; neighbors sue each other. And nobody really wins. The homeowners may recover money, although that is certainly not guaranteed, because shady or financially strapped contractors are notoriously difficult to collect from.

The money lost is not necessarily the worst of it. A remodeling gone wrong also wastes time, escalates stress, destroys peace of mind, and prolongs the mess. Homeowners in the midst of a mistake can lose the full use of their house for a time. If the problem is bad enough, they could lose their home completely.

Here's some more semi-bad news. Approximately 25 million homeowners will undertake some kind of home-improvement project every year, according to the Harvard-MIT Joint Center for Housing Studies. But the number of professional remodeling contractors in the country, estimated at between 600,000 and 800,000, cannot meet the current demand for their services. This explains why some professionals are booked many months in advance. And it also means that some-not-so-professional contractors are out there taking up the slack, looking for work, and getting it. In this sort of climate, it is extremely important to be careful and thorough when checking out people to oversee your renovation project.

MUTUAL RESPONSIBILITY FOR A POSITIVE OUTCOME

Part of the reason that the very word *contractor* strikes fear in so many people's hearts is, of course, the many horror stories we've heard, and believed. There is no doubt that some people have been horribly taken advantage of by unscrupulous contractors. But in fact most of these stories reveal mistakes by both parties: contractors who somehow mismanaged the job, and homeowners who didn't take proper and sensible steps to protect themselves.

Another reason for contractor-phobia is unrealistic expectations. We think the job should go absolutely smoothly; and when it doesn't, we inevitably blame the contractor for being incompetent or even dishonest. But things do go wrong on a remodeling project. Always. "It's normal," says N'ann Harp, president of Smart Consumer Services, a consumer education and assistance organization based in Arlington, Virginia.

"Because there are literally so *many* details; so many questions to ask; so many decisions to make, one leading to and affecting another; so much coordination required—even under the best of circumstances it is likely that something will fall through a crack and have to be revisited, fixed, or modified," says Harp. "To *not* expect something to go wrong is to have unrealistic expectations."

Yes, there are a few unscrupulous people who want to take your money and leave town. Don't worry; we'll help you learn how to spot them. And there are others who do shoddy work and try to cheat you on the job; we'll show you how to avoid them too. But most contractors are honest and hardworking. Confusion, haste, and careless business practices are the causes of problems more often than not. For example, contractors may be a lot better at building than they are at business. They may make poor decisions, such as taking on too many jobs, and fall behind on their bills; and then make more poor decisions, such as leaving one job not quite finished so that they can start another one. This all-too-familiar practice is the basis for many remodeling nightmares. And, yes, we'll show you how to protect yourself against *that* too.

Confusion can reign on both sides. According to N'ann Harp, "Remodeling contractors and homeowners share equally in the responsibility to eliminate as much confusion and as many unanswered questions as possible before the job begins." But homeowners often fall down on the job too. Harp says they make two mistakes: failing to check all references, and failing to get a legal review of the contract. And, we hasten to add, failing even to insist on a contract in the first place is another very common mistake.

To help you avoid this mistake and many others, we'll take you step-by-step through the contractor-selection process, advising you about where to find good prospects, how to interview them and thoroughly check their references, and how to recognize problem candidates. We'll also walk you through the bidding phase and provide you with detailed information about preparing a contract that protects both you and the contractor. And we'll clue you in on what you can expect once the job is under way and contractor and crew have become part of your household.

So the next time you hear a remodeling horror story, don't automatically assume something like it, or worse, could happen to you. After greeting the story with the appropriate oooh's and aaah's, ask the teller of the tale some gentle questions (be careful here; some people relish their negative experiences and don't want to know how they could have avoided them). Your questions could be: did you check the guy out, ask for references, draw up a contract, have the contract reviewed? Chances are, the answers will be no, no, no, and no.

After having thus defused the power of contractor horror stories, you can read this book and begin planning your renovation, knowing that you have a lot of control over the process and the ability to create a smooth remodeling experience with a successful outcome.

Before You Begin

A Typical Timeline

Every remodeling project is different. Some present more problems than others; some move quickly, while others seem to linger. Still, in every project schedule there are certain universal steps that ensure a successful outcome. Here, courtesy of the National Association of the Remodeling Industry (NARI), is an outline of those steps.

STEP 1: PREPLANNING

- Make the decision to remodel.
- Assess the scope and probable cost of the project.
- Determine which professionals you will hire—interior designer, architect, contractor.
- Gather names of architects, designers, and contractors to interview. Check all references.

STEP 2:
DESIGN AND AGREEMENT
(IF USING DESIGN OR ARCHITECTURAL SERVICES)

- Ask design professional to present preliminary drawings and project specifications.
- Review and discuss options.
- Talk to previous customers to be sure you and the design professional are a good fit.
- Have the design agreement reviewed by legal counsel before signing it.

STEP 3:
THE FIRST MEETING WITH THE CONTRACTOR

- Set up interviews with at least three contractors and do a walk-through of project.
- Check all references thoroughly.
- Ask for bids from the contractors who remain on your list.
- Assess bids carefully, making adjustments if necessary.
- Select the contractor you want to work with, draw up a contract, and make the first payment.

STEP 4:
PROJECT DEVELOPMENT

- Prepare plans and construction drawings.
- Review drawings to clarify any misunderstandings of the scope or intent of the project.
- Work with contractor or design professional on selecting products and materials.
- Have the contract reviewed by legal counsel before signing it.
- After signing contract, wait three days for Right of Recision period to pass. (See page 58.)
- Make the first payment.

STEP 5:
PRECONSTRUCTION

- Request that contractor obtain permits.
- Meet with contractor and design professionals, if any, to create project schedule.
- Give final product approval and request that products be ordered.
- Finalize project schedule.

STEP 6:
PREPARING THE HOUSE AND THE FAMILY

- Establish house rules and off-limits areas for contractor and crew.
- Use plastic sheeting to protect rooms and furniture from dust and debris.
- Remove household items from construction areas.
- Call a family meeting to discuss the project and its impact on home life.

STEP 7:
THE PROJECT

- Demolition.
- Framing.
- Roughing-in of plumbing, electrical circuits, heating ducts.
- Flooring.
- Wallboard.
- Cabinetry.
- Make installment payments.
- Fixtures and appliances installed.
- Finishing touches.

STEP 8:
THE FINAL DAYS

- Create punch list (or a “loose-ends” list).
- Schedule final inspection.
- Give final approval (after loose ends are taken care of).
- Make final payment.

STEP 9:
AFTER THE DUST SETTLES

- Review maintenance schedules for new products.
- Move in and enjoy!

one

Who Do You Need to Hire?

If you're about to embark on a major renovation, or even a minor remodeling project, you're going to need professional help—an architect perhaps, an interior designer maybe, but a contractor, almost certainly.

Before we guide you through the maze of finding the right contractor for your job, let's define a few terms.

The word *contractor* (which Webster defines as one who enters into a binding agreement "to perform work or provide supplies on a large scale") means, for our purposes, a general contractor, sometimes called a remodeler or builder.

There may be upwards of a million remodeling contractors in this country. That's a rough figure, because in an industry that is not strictly regulated it is impossible to count all the people who advertise themselves as contractors and do remodeling for hire. The National Association of the Remodeling Industry (NARI) claims about 6,000 members; the Remodelors' Council arm of the National Association of Home Builders also lists 6,000; and ImproveNet, an online service that advises consumers about many aspects of remodeling, claims to have 600,000 contractors on file. There are plenty more contractors who are honest and solid but have not joined any of these organizations.

You may also hear the word *contractor* used to describe specialists who work in such specific areas of home improvement as roofing, plumbing, electricity, heating and air conditioning, painting. These people are experts only in the work suggested by their titles. Although a roofing contractor or plumbing contractor may play an important part in your project, he or she will probably be hired by your general contractor and not by you.

A general contractor is the man or woman you'll be looking for. (Yes, *of course,* there are women contractors. According to a recent report from the Census Bureau, about 230,000 construction businesses are owned by women, and 134,000 women were earning a living as residential general contractors in 1996.)

WHAT DOES A CONTRACTOR DO?

Experienced in all aspects of the building/home-improvement field, a general contractor will be in charge of your entire project, setting the schedule; ordering materials, supplies, appliances, fixtures, and so on, and coordinating their arrival on the site; hiring and overseeing the work of individual tradespeople (called subcontractors, or subs). Contractors also work with architects and designers, cope with the inevitable changes and surprises that occur, obtain necessary permits, and arrange for inspections of the job.

N'ann Harp is president of Smart Consumer Services, an Arlington, Virginia, consumer education and assistance organization that specializes in homeowner-contractor relationships. She sums up the contractor's job in this way: "The transformation of a set of blueprints on paper into a living structure in physical space requires the coordinated teamwork of up to a dozen, or more, small groups of specialty contractors and suppliers. It is the general contractor's job to orchestrate the hundreds of details involved in making that transition happen, from sourcing and pricing all of the materials to scheduling, overseeing and paying workers. The job calls for the ability to process and retain large volumes of details while simultaneously exercising firmness, patience and judgment, and using interpersonal management skills with the work crews."

For your project, a contractor will probably do all the following tasks:

- Obtain building permits.
- Provide labor and construction materials.
- Coordinate insurance coverage and workers' compensation for the subs.
- Hire and pay all subcontractors and materials suppliers.
- Perform demolition and haul away debris.
- Schedule the subcontractors and keep them on track.
- Manage the overall construction schedule.
- Manage and coordinate the construction work.
- Meet regularly with homeowner and designer and/or architect to answer questions, solve problems, discuss changes.
- Arrange building inspections as required in your locality.

"I'm a huge fan of experienced, qualified builders and remodeling contractors," says N'ann Harp. "At their best they are adept at handling untold complexities of retrofitting, scheduling, and logistical nightmares, and can they be amazingly imaginative problem solvers."

Contracting businesses come in three sizes: small, medium, and large. Some large contracting firms have many major jobs going on at once and several permanent crews working; this sort of operation also has an office staff and may employ architects, designers, as well as carpenters and other specialists. The person you interview may be a sales agent for the company who will probably not perform any actual work on your job but will be on-site daily as a project manager, assigned to the logistical management of individual jobs. (Note: If you go with a large firm, at some point in the process you should also interview the general contractor who will be assigned to your job.) Advantages of a large operation include speed (large crews allow them to do the job fast) and financial solvency (they won't leave you in the lurch to go to another job because their money has run out).

Most contracting businesses, however, are medium or small and may offer lower prices because they have less overhead than the larger firms. One type of firm is not necessarily better than another; your choice will depend on your needs, budget, and the scope of your project.

A mid-size general contractor probably has an office in his home, handles one or two projects at once, employs a stable of regular subcontractors, and is on the job every day supervising those subs. Although this type of contractor may sometimes handle new-construction projects, chances are that he or she specializes in renovations. A small operation generally consists of just one person (often with a helper) who does his own carpentry, uses subs only occasionally, and works on small projects. If you're planning a major renovation costing from $50,000 to $100,000 or more, the mid-size operation may be the one to aim for. The large firm may find a project of that size too small to bid on; and the small operation, which usually consists of one person who hires an occasional helper or two, probably prefers not to work with the large crew needed for a major project.

Some contractors are part of design-build companies. These businesses generally employ an architect, architectural designer, or engineer who will create a design for your addition or renovation, and a contractor who will build it. They perform every phase of a project from the initial design all the way through to completion. Kenneth Skowronski runs such a firm, KS Remodelers in Franklin, Wisconsin. "We're the general contractors," says Skowronski. "Working with the clients, we develop the concept, our on-staff architectural designer produces the design and drawings, and we do the construction of the project down to landscaping, wallpapering, and painting. One of my lead carpenters oversees the job, and I occasionally stop by to see how it's going. It works well for the homeowners because they don't have three separate people charging them money."

N'ann Harp points out that in most areas, once a general contractor—large or small—has been in business for a while, he or she develops a preferred range of job size and complexity with which he feels comfortable. "Contractors," she adds, will initially tell you that they 'do everything,' but if you ask them, they'll usually let you know what sorts of jobs they really like to do. This is important to the homeowner because what a contractor likes to do is what he or she will be most experienced at and will do best."

Hire the Right Person for the Job

Not all remodeling jobs require a general contractor. Some projects are best handled by a specialist of some kind, a person who does minor maintenance and repairs, or even by homeowners themselves.

REMODELING PROJECT	THE BEST PERSON TO DO IT
• Paint or wallpaper the interior; lay a vinyl-tile floor.	Minor-maintenance person or handyman; these projects are also appropriate for a skilled do-it-yourself homeowner.
• Install paneling.	Carpenter; homeowner.
• Weatherproof windows and doors.	Handyman; homeowner.
• Replace windows.	Window contractor; store installer, carpenter.
• Paint the exterior.	Painting contractor.
• Lay a wood, ceramic-tile, or sheet vinyl floor.	Flooring contractor; store installer; carpenter.
• Install kitchen cabinets.	Handyman or homeowner can handle demolition; but store installer or carpenter should install.
• Replace old bath fixtures.	Plumber.
• Update wiring.	Electrician.
• Add a new roof.	Roofing contractor.
• Install siding.	Siding contractor or carpenter.
• Build a deck.	Carpenter; skilled homeowner.
• Finish the basement.	Carpenter; skilled homeowner.
• Add dormers to the second story.	General contractor; carpenter.
• Convert attic to living space.	General contractor.
• Add a room, a wing, or a second story.	General contractor.

HOW DID CONTRACTORS GET SUCH A BAD REPUTATION?

As you can see from the description of all the work that a general contractor must do and oversee, this is not a job for the faint of heart or the inexperienced. And yet many contractors have had no schooling or training in the carpentry or building trades nor are they required to have such training by any of the states that license them.

Only 30 states demand that general contractors be licensed to do residential work, and only 24 states require that applicants for a license take a qualifying exam. The qualifying exams typically test the applicants' overall business knowledge and specific knowledge of their particular trade as well as their understanding of state contracting laws and regulations. And licensing requirements notwithstanding, many contractors operate without any license or official approval at all. So, in an industry that is so loosely regulated, a questionable and dishonest element is inevitably present. (For tips on how to spot out-and-out con artists, who may have no home-improvement knowledge at all, see Chapter 2.)

A larger and more troublesome category is the contractors who are honest and aboveboard but don't do good work. Maybe they have full-time jobs but are moonlighting as contractors because they've been able to pull off a few simple projects at home. Maybe they are pretty good at rough carpentry and have had success with a few straightforward projects—such as building a deck or a set of exterior stairs—and on the strength of that, they buy a pickup truck, invest in a few tools, and call themselves contractors. They don't intend to rip anyone off, but they always cause trouble, and the jobs they bungle add to the jaundiced view that many people take of contractors in general.

Still another category, probably the largest of the questionable contractor group, is made up of those people whose work is good but whose business management is bad. That is, they are poor at estimating the probable cost of a job; careless about setting aside money for insurance, tool maintenance, gas, taxes, and so forth; disorganized when scheduling the work; and ineffective at supervising their subs and getting a good day's work out of them. These are the people who cause the most trouble for their customers and for themselves. We'll show you how to identify and avoid them. Both careless and larcenous contractors can be unearthed by careful checking and kept in line with wisely drawn-up documents.

WILL YOU NEED A DESIGN CONSULTANT?

It's a given that you will need a general contractor to oversee and manage any significant home-improvement project. But unless your contractor is part of a large firm that also employs its own designers and architects, you may also need to hire the help and expertise of other professionals.

Although it is possible to go right from idea to contractor and skip an architect or interior designer entirely, you should at least consider hiring one of these professionals.

"With rare exceptions, g.c.'s don't provide more than rudimentary design services," says N'ann Harp. "A general contractor who has been in business for many years will have picked up much practical knowledge of floor plans and ways of utilizing space. If you ask him or her for design input, however, the inclination will be to suggest what is easiest to build. Keep in mind, you're the one who will have to live with what is built, not the builder. Contractors, as a rule, prefer their work very simple or spelled out in detailed drawings."

Susan Aiello, a designer with Interior Design Solutions in New York and a member of the American Society of Interior Designers (ASID) believes large jobs need professional guidance. "You should always hire a design professional if you're doing major construction," she says. "Failure to do so is penny-wise and pound-foolish.

The type of assistance you need depends on the scope of the project. "If you are planning a substantial addition, an architect would have the skill to harmonize it with the existing exterior of your house and with the landscape. An interior designer is probably the best person to work with if you want a fabulous kitchen or a luxurious master suite but don't plan to change the exterior of your house."

For a straightforward job that requires no structural alteration or major changes—such as sprucing up a dated bathroom or adding a sunroom to the rear of your house—you and the contractor can probably go it alone. But the larger the project, and the more money you will spend on it, the greater the possibility that you will need design help. One rule of thumb that some people in the field use maintains that if your renovation will cost more than $25,000, it's wise to hire an experienced design consultant of some kind.

Although architects and contractors have been known to behave like adversaries, some contractors like working with architects, even prefer it. Mike Vella of Vella Interiors in Long Island City, New York, says: "If we have a job that involves a lot of historical detail, say, renovating a brownstone, I like working with an architect. It's better for the client too. In a case like this, I prefer working with people who know what they're doing. On some of my jobs, I need to take direction from a design professional. And, in fact, I won't take a major, high-end job unless there is an interior designer or architect involved.

"If you want to save money, don't skimp here," is the advice Vella offers homeowners contemplating a big project. "The planning stage, for which an architect is invaluable, is the most important part," he says, "and if it's done poorly, a three-month job can take six months."

It's true that the services of a design professional will add to the cost of the project, which is why many people balk at hiring one; but keep in mind that a well-conceived, architect-designed project can probably be built more economically than something you and your contractor slap together. The extra money may be a very wise investment. You be the judge.

Architects

Trained to consider all aspects of your project, architects typically have earned a bachelor of science degree in architecture, apprenticed for three or more years, and passed a licensing test administered by the state in which they practice. Architects are trained to have a working knowledge of the complete building environment: they've studied engineering and site planning as well as structural and aesthetic design. They can do any or all of the following things for you:

- Create a design for your project.
- Prepare construction drawings.
- Make suggestions about windows, doors, trim, moldings, cabinets, and other design elements.
- Ensure that your project will meet local building codes.
- Help estimate costs.
- Oversee the bidding/negotiation process with contractors.
- Oversee the project during construction, verifying that the project is indeed being executed according to the plans and specifications.

- Be available to circumvent problems or to solve them as they arise.

Five Projects in Search of an Architect

According to John Ingersoll, a writer who specializes in architecture, design, and home improvement, substantial renovations cry out for the experience and good design sense of an architect:

- Adding an entire wing to an existing house calls for extensive design know-how. The ability to blend new and existing facades and to weave additional rooms into a coherent overall plan with a logical, streamlined traffic flow is critical.
- Putting a second story on a ranch house demands the same skill for matching new and old exteriors and interiors.
- Changing the architectural style of a house requires a sure sense of architectural history and the ability to make the changes economically.
- Accurately restoring or altering a historic house entails not only strong design skills, but also a thorough knowledge of period architecture.
- "Staying within a period style definitely calls for the aesthetic training of an architect," says Mark Downey, an architect based in Lake Forest, Illinois. "I think it would be difficult for a contractor who is not an architect to achieve that kind of design quality, considering all the other pressures of running a construction company."
- Gutting an old house and starting over takes as much design aptitude as planning a new house, maybe more.

Steven House, who with his wife, Cathi, owns and operates House + House Architects in San Francisco, thinks large projects require an architect. "My wife and I have been doing remodeling projects of all shapes and sizes for 17 years. We've found that architects are invaluable in bringing creativity to the project and advising clients how to put their money to best use. We know the value of important products, such as windows, appliances, fixtures, cabinets, and surfaces, and we can help clients choose materials and finishes that are durable as well as beautiful."

Another important role of an architect is to act as an intermediary, says House. "Because we have both knowledge and experience, we can fairly assess problems that may come up. We are constantly resolving issues and keeping them from snowballing," he says. For example, in makeovers of older homes there are often unpleasant surprises when walls are opened to add insulation or bring wiring up to code. The contractor may discover termite damage, say, or venerable old plaster walls may start crumbling. Both surprises require extra work that can cost the client extra money. "The clients are in a difficult position," says House. "They don't want to create an adversarial situation with the contractor, but on the other hand they don't want to spend money unnecessarily. An architect can function as an intermediary here and help determine what is fair, what the contractor should have known, what the homeowners must take responsibility for."

A change order, which can cause tense moments on a renovation project, is usually executed more smoothly with an architect on the job, House claims. If, in the midst of the project, the homeowners change their minds about the layout of the bathroom and want to move a sink and add a shower, it could greatly increase the cost that was originally estimated for that room. "There may be conditions, such as location of a beam, that raise the cost," says House. "But meanwhile, the owners are furious and so is the contractor. An architect can step into the fray and explain to the owners why this change will cost so much or in what ways it will compromise the timing of the project. And sometimes," he adds, "we're able to convince the clients to let that particular change go.

"My wife and I have taken care to become knowledgeable about interior design as well as architecture, so that on our projects the clients don't really need a separate designer," House continues. "We create the space and then help owners select furniture, cabinets, surfacing materials, carpets, and so forth. Not all architects are good at this or interested in doing it. In that case, I would suggest that the homeowners bring an interior designer on board. And he or she should come in at the beginning. It's frustrating to have one person stop and the other begin. Ideally, all the professionals will work together at once—landscape designer, architect, structural engineer, interior designer, lighting designer, or whatever the project requires. If all of these professionals define the scope of the work initially, there will be no overlap to cause problems later."

If your budget is really tight but you want design help on your renovation nonetheless, you may be able to work with an architect on a short-term basis, whereby he or she would create a design and produce plans and working drawings, and you and your contractor would take it from there. Some architects are willing to do this; some are not. Shop around until you find one who will cooperate.

Whether or not you hire an architect for design help, plans for your project must be signed by an accredited architect before your local building department will approve them. More on this in Chapter 4.

Interior designers

Your project may also benefit from the services of an interior designer. These professionals have been trained to deal with all design-related, nonload-bearing structural issues. Most hold a degree in interior design and many are accredited by a professional organization such as the American Society of Interior Designers (ASID) or the International Interior Design Association (IIDA). In addition, more than 20 states require interior designers to be licensed. Such a background enables an interior designer to do space planning and to prepare working drawings and specifications for nonload-bearing interior construction. He or she can advise you not only on furniture, finishes, and fabrics but also on flooring, plumbing fixtures, appliance layout, and such technical issues as fabric flammability and product performance.

If you're building an addition with, say, a family room on the ground floor and the master suite above it, an interior designer can help you coordinate the character of the new rooms with the rest of your house (especially important if your new family room opens to your kitchen, as many do these days), And he or she can help you select everything from furniture to accessories and other decorative details. Because designers are so familiar with the field of home furnishings and have access to showrooms, they can greatly expand your decorative options.

Don't confuse interior designers with decorators. Although a decorator can help with color schemes and furniture selection—many department and furniture stores employ them—an interior designer does more. Or as Susan Aiello, ASID, puts it on her web page (www.idsny.com): "Interior design is more than just decoration. We specialize in uncovering the hidden potential of a house or apartment that is in less than mint condition and transforming it to meet your needs. And a designer may help you communicate with your architect, who may not be interested in such details as window treatments or tile or cabinet hardware.

"Of course you can decorate your own home," says Aiello, whose firm, Interior Design Solutions, is located in New York, "You can also sew your own clothes and cut your own hair. The best results in any of these endeavors, however, are usually achieved by people with a certain level of skill and experience."

Many homeowners *do* have that skill and at least some of the experience. If you're confident about your own design abilities and you can clearly visualize your new rooms, fine. But don't reject the idea of design help strictly on the basis of money. Some designers respond to a tight budget by agreeing to undertake a consultation or two and some sketches, leaving you to take it on your own from there but providing invaluable help for you to find your way.

"One way to deal with a limited budget is to hire a designer to create a design and produce working drawings and then turn the job over to you," says Rick Shaver of Shaver Melahn in New York. "It's not difficult to find a designer who is willing to take on a project like this. Sometimes this kind of short-term arrangement also includes a furniture layout and specifications for finishes, fabrics, and colors. Although I prefer to see a design project all the way through, I have worked like this on occasion," says Shaver, an interior designer and creator of a new line of moderne-inspired furniture.

"You need to have a really good detailed plan so that everyone knows where they are going," says Shaver, "and for that you need a professional. Most people don't have the vision and the wherewithal to deal with a major remodel. Otherwise, you could in effect be trusting your project to the aesthetics of your contractor." In some cases, that might work out but generally speaking contractors do not make aesthetic decisions and do not necessarily know what something is going to look like. They are not trained to handle the aesthetics of a job.

Shaver notes that a designer can also be valuable as an advocate, a mediator between you and the contractor. "Although the contract exists between the client and the contractor, on my jobs payment will not be made until I sign off on the work, which means that it has to meet my standards. Did they follow the design properly? Is the work—such as spackling, taping, painting, carpentry—of good quality? With someone like me on the job, the clients can be friends with everybody. If they have a problem, they call me and then it becomes my responsibility to straighten it out."

Should you decide to hire both an architect and an interior designer, it is best to get both of them on board early in the project so that they can work together from the beginning. And draw up contracts with them, as well as the one you will enter into with your general contractor.

Heads Up: Get It in Writing

It is important that you enter into a complete, detailed, written contract with *every* design professional that you hire for your home-improvement project. The contract should clearly spell out the following: the work that the professional will perform, your responsibilities, time schedule, amount of fee, and payment schedule. A good contract minimizes misunderstandings and confusion, and goes a long way toward preventing problems. Don't even think of working without one.

Kitchen and bath designers

The National Kitchen and Bath Association (NKBA) offers courses of study that certify its members as kitchen and bath designers. The initials CKD or CBD (Certified Kitchen Designer or Certified Bath Designer) after a person's name indicates that he or she has completed courses in numerous aspects of kitchen or bath planning, passed a rigorous exam, amassed a minimum of seven years' experience, and is equipped to handle all aspects of the creation or remodeling of a kitchen or bath. "All aspects" include space and layout planning; installation of cabinets, appliances, and other equipment; and selection of materials, finishes, and decorative details. Certified designers conduct design consultations, prepare blueprints and specifications for the job, and are experienced at working with contractors and subcontractors and at scheduling projects. These kitchen and bath specialists are often also dealers who maintain showrooms that display products and complete designs.

Where to find a design consultant

Good places to look for an architect are among friends or neighbors who have extensively remodeled a house in the past year or so, in the home section of your local newspaper, and in consumer-shelter magazines (those that cover architecture, decorating, remodeling). Another good source is a local remodeling project that catches your eye. Whether the project is recently completed or in process, ask the owners for the name of their architect. Don't worry if you don't know the owners. Just ring the bell, tell them you admire the work that's been done on their house, and ask who did it. It sounds bold (and it is) but people are usually very pleased to be asked.

Another way to find an architect is to visit the local chapter of the American Institute of Architects. AIA chapters often have scrapbooks of residential work designed by members. Take down the names, addresses, and phone numbers of the architects whose work you like. If there is no AIA office in your town or nearby, call the national office, which is listed in Chapter 7, to find the chapter closest to you.

Once you have a list of architects' names, run them past your local building department. The department will not make recommendations, but you will probably be able to pick up some clues as to which architects are known and respected and which are not. You don't want to hire an architect who is *persona non grata* with the building department, no matter how brilliant his or her work may be.

To find an interior designer or a kitchen/bath specialist, use the same strategy. When you see a professionally designed room that appeals to you, whether in a newspaper, magazine, or a friend's or neighbor's home, find out who did it. If the designer is local, or at least accessible, you may have a candidate. For further referrals you might contact the nearest office of the American Society of Interior Designers. This professional organization represents about 20,000 designer-practitioner members who have satisfied educational and field work requirements and passed an extensive exam. If there is no

ASID office near you, contact the Washington, D.C., headquarters at the address listed in Chapter 7. If you cannot find a Certified Kitchen Designer in your area, call the NKBA, which is also listed in Chapter 7.

If possible, gather names so that you have several candidates. Then call, initially to find out whether the architect or designer works on projects like yours, has the time to give to your job, has any interest in doing it. If the answers to those preliminary questions are affirmative, set up an interview.

When selecting a design professional, here are some questions to ask and things to look for:

- How long have you been in business? (The longer, the better.)
- What percentage of your work is in residential design, particularly renovations? (You're looking for about 50 percent. Avoid those who work predominantly in commercial projects. You don't want your renovation to look like a bank lobby, after all.)
- Do you have any experience with projects similar to mine?
- Performance. Get this information from talking to previous clients. (Restrict your inquiries to those who are in your general economic category. The Joneses may be great friends of yours, but if they could afford to pay, say, $400,000 for their renovation and you can't, you may be barking up the wrong design tree.)
- Communication and responsiveness. (Look for someone who is relaxed and seems to enjoy interacting with people.)
- Fee structure. (Ask what the architect or designer would expect fees to be for your project. Explain your budget concerns to the candidate. (We provide some suggestions on how to establish a budget in Chapter 2.) If he or she is unwilling to work within economic restraints, or unwilling to commit to a figure within 15 percent of your budget, look further. Ballpark figures can seriously strain your budget.)
- A common vision. (This is a little tricky. Basically, a good

architect or designer should give you what you want, but you don't want to tie his or her hands and squelch the design creativity for which you hired a professional in the first place. Take an active role and communicate constantly about the way you want the renovation to look and to function. Most professionals will work with you toward the best solution. But if your ideas, and your budget concerns, aren't ultimately reflected in the proposed plans, you should probably look elsewhere. Why end up living in an expensive monument to someone else's ego?)

Heads Up: Stick Up For Your Style

Contrary to conventional belief, you and your interior designer do not have to have the same taste. Study pictures of his or her completed projects, yes; but don't be disappointed if they're not "you." As long as the work looks professional and well-executed, and the photos demonstrate design proficiency in several styles, you know the designer will be able to respond to your particular needs and tastes. Do beware, however, if the designer's style departs radically from your own in all cases or if all the rooms in the portfolio seem to be making the same strong stylistic statement. Unless you love that particular statement, you may have trouble getting this designer to respond to your preferences.

GOING IT ALONE—JUST YOU AND YOUR CONTRACTOR

Although there are lots of good reasons to hire an architect for a major renovation, many people don't do it. In fact, it is very common in suburban and rural areas of the country, where design professionals may be in short supply, for homeowners to turn their major remodeling projects over to a contractor and dispense with the services of an architect or designer altogether.

How do these projects turn out? Not too well, probably, not unless there is what N'ann Harp calls a "hidden designer" on the job. "Just because a contractor is skillful at using tools doesn't mean he or she is a designer," says Harp, who is a former contractor herself. "If one of these contractor-only jobs comes off, it's probably because the homeowner has design skills. Or maybe the contractor does." If nobody has design skills, the job could go very badly or the result could look awful, or both.

Debbie Lumia was the "hidden designer" on the building of her house in Delhi, New York. When Debbie and her husband, Frank, decided to build about 10 years ago, there was no question of hiring an architect. Debbie had absolute confidence that she would be able to create an aesthetically pleasing, well-proportioned house that would suit their needs and tastes. How did she come to be so confident? "I was very clear about what I wanted," she says, "and I had some skill at design, drawing, and conceptualizing space. My background in layout and graphic design helped with that."

The design evolved over time, and finally Debbie drew up plans for the project, which were good enough to impress the contractor she had hired. "I read several books and taught myself how to draw plans," she says. "And being a slightly obsessive person, I included every detail. I even drew in the windows down to one-sixteenth of an inch." After her drawings were approved by an engineer, they were used by the contractor to build the house.

Charlotte Hill, another intrepid homeowner, recently completed an expansion of her house in Delancey, New York, without any professional design help. Hill had the ideas, and her contractor of choice, Bruce Commandeur, translated them into reality. The project, a refurbished kitchen and dining room and a new family room, went off without any serious hitches.

In this case, both parties were "hidden designers." Hill knew exactly what she wanted, could visualize it, and was very clear about explaining it. Commandeur, who studied photography and layout before he went into the construction game, was easily able to put her ideas onto paper as working drawings and to offer input on visually balancing the components of the design. He found Hill's certainty very helpful. "Most people don't know what they want," he says. "Together we measured the project out on graph paper, drew in the location of walls, doors, and windows, considered furniture placement, and went back and forth until we had something workable."

Success stories for large projects like these without a professional designer on board are relatively rare. Debbie Lumia succeeded because she had skill, patience, and a background in design, and was able to give a great deal of time and attention to the creation of her house. Charlotte Hill's renovation was relatively simple. The one-story, shed-roof structure that encloses her new living space was added to the rear elevation, where it is not noticeable that the roofline does not match the Colonial-style architecture of the existing house. Had contractor Bruce Commandeur been asked to seamlessly blend old and new rooflines, he might have been out of his element.

Books and magazines that feature home-remodeling projects can inspire the hidden designer in contractor-only jobs. Since each makeover of an existing house is unique, it is not possible to come up with specific predesigned construction drawings. Even so, a perusal of typical remodelings can be a great help. Jerry Axelrod, a Commack, New York, architect, has written a book that addresses this situation. His *Plans for Adding On and Remodeling* (New York: McGraw-Hill, 2000) focuses on design ideas and approaches that help homeowners plan their projects themselves.

THE DO-IT-YOURSELF DILEMMA

General contractors—good ones, that is—are experts in the home-improvement field. You probably aren't.

Houses are made up of intricate systems and subsystems, all of which are connected. Electric wires and plumbing pipes snake their way through walls, floors and ceilings, which are covered by wallboard, plaster, wood, or other materials. The walls, floors, and ceilings are supported by posts and beams; sills support these large framing members; and the whole structure rests on a foundation. No home-improvement professional is an expert on all these things, but a good general contractor knows a certain amount about all of them. More important, he or she knows the people who *are* experts on these various systems. A g.c. also knows how to schedule a project that will remodel some, or all, of the systems.

Even if you do know a lot about carpentry and certain of the building trades, such as carpentry and plumbing, you probably don't have the time to tackle your own renovation. (Read about exceptions to this below.) This is not to say that you shouldn't become very familiar with these topics in order to oversee your project; you definitely should, as we will discuss in Chapter 2. But trusting the job to the right professional is generally considered the best way to go.

Remodeling projects cost money, a lot of money, which is probably one reason that homeowners get fired up about doing some of the work themselves. We'll save money, they think. And it's true that you can save perhaps a couple of thousand dollars, or approximately 10 to 20 percent of construction costs, if you are willing to handle some simple tasks such as demolition, daily clean-up, painting, wallpapering, even laying a simple vinyl tile floor.

But in no case should you consider taking this approach unless you possess both skill and experience. Sure, painting is easy, but if you're not really good at it, or don't schedule enough time to do it properly, the surfaces you paint will look shoddy and will present a dreadful contrast to the rest of your professionally executed project. A recent client of Steven House made this mistake. "He was on a super-tight budget," House recalls, "and as we do from time to time, we took some simple jobs out of the budget so he could handle them himself. He took three weeks away from his job so that he could paint his rooms, but did it so sloppily that he had to hire a professional to redo it." It took the client three weeks away from his job to do the painting badly; it took a professional three or four days to do it well. Result: a lot of wasted time and money.

Katie and Gene Hamilton, who have made a career of doing home-improvement projects themselves and writing a weekly column and several books about it, claim to have done "just about every job imaginable remodeling and repairing houses." Consummate do-it-yourselfers, they learned an important lesson in their 30 years of updating houses: sometimes you can save time and money by *not* doing it yourself and hiring a contractor instead. They discovered this after sprucing up their fourth house and getting ready to refinish the old hardwood floors, a job they had done many times but with only so-so results. They always spent a bundle renting the equipment, they expended a lot of energy doing the work, and after all that, the floors never looked really good. After getting a reasonable estimate from a local floor sander, they split up the job: the pro did the sanding, and the Hamiltons applied the finish. The result was beautiful, far better than their own previous efforts. Even taking into account the expert's fee, the Hamiltons were still on budget and a week ahead of schedule.

In one of their books, *Do It Yourself...Or not?* (New York: Berkley Books, 1996), the Hamiltons share two other important lessons they have learned: (1), DIY home-improvement projects often take longer and cost more money than planned; (2), you can't assume you're going to save any money at all until you factor in the cost of purchasing and transporting materials and equipment.

But the appeal of the do-it-yourself experience is about more than saving money. It's also about the romance of it all. We'll have a great feeling of joy and satisfaction, we think, after having built our own house or renovated our kitchen. In a country settled by rugged individuals, many of us have nourished a strong leaning to self-reliance in our own lives. Early Americans built their houses themselves, and so, we reason, should we. A very commendable attitude in many ways, but it's still best to leave certain tasks to those who have the expertise.

Contractor Mike Vella discourages DIY impulses in his customers for that very reason. "I'm very much against it as long as I'm on the project," he says "I don't like working with someone I haven't hired myself. Also, a sloppy do-it-yourself job messes up the coordination and scheduling of the project and could even make our work look bad."

On the subject of do-it-yourself, architects Steven and Cathi House have a definite philosophy. Work at your own job, at which *you* are an expert, they advise clients, and make the money to pay another kind of expert to do your remodeling.

But if you think you are qualified—our self-quiz on page 21 will help you make that judgment—by all means approach your general contractor about working with him or her on this basis. Make sure the project is carefully coordinated to allow for your contribution and that responsibilities and liabilities are clearly defined in your contract. Any work you do should be on your own time and should not involve the contractor.

"Sometimes," says designer Rick Shaver, "I take a real tight-budget job and work it out with the clients that they will do some of the work themselves, although basically that decision is between them and the contractor, who has to handle the schedule and make everything dovetail. In these situations you have to lay out everybody's duties beforehand, and you have to be very clear about what needs to be done and who does it. For example, if the homeowner is going to save some money by painting the walls, you need to know whether he or she is just going to paint or also prepare the walls and provide all of the equipment.

"This do-it-yourself approach can be managed but usually I'm reluctant," Shaver continues. "I find that it's really better to have one guiding hand on a project. Fewer mistakes are made and the costs of those mistakes are minimized."

What's Your DIY IQ?

Before you make definite plans to take on some of the tasks in your home renovation, take this quiz to assess your do-it-yourself abilities. If you answer "no" to more than five of these questions, doing it yourself is probably not a reasonable option for you.

- Do you have a clear idea of what you want your project to look like?
- Do you enjoy physical work?
- Are you patient and persistent?
- Have you ever undertaken a project like this before?
- Do you understand all of the safety issues associated with this project?
- Do you know what materials, tools, and equipment you will need? Do you own the tools? If not, can you easily rent them?
- Do you know where to get the necessary materials?
- Do you have reasonable skill at painting, staining, wallpapering, basic carpentry, laying tile? If not, forget it. This is not the time for on-the-job training.
- Are you familiar with applicable building codes and permit requirements for your renovation?
- Do you really have the time to give to the project? Failure to complete your DIY tasks on time could slow down or even shut down the entire remodeling process.
- Will you need help with the DIY project? If so, who will help you? Does that person have the time and skills necessary?
- Are you willing to do your homework and prepare for each task? Even painting, the easiest of all DIY jobs, requires some preparation: a step stool or ladder, good brushes and rollers, the right paint, plenty of drop cloths.

Remember, your home is an important financial investment, for most of us the largest we will ever make. Improving that home with a well-planned remodeling can greatly enhance the investment, as we will see in a later chapter. Conversely, messing it up with a poorly executed do-it-yourself project can send that investment right down the tubes, its resale value seriously compromised and its "curb appeal" diminished. You may be able to sell your house with its shoddily remodeled kitchen in which the cabinets aren't level and the vinyl tile floor is curling but you won't get the money for it that a good kitchen will bring in.

ACTING AS YOUR OWN GENERAL CONTRACTOR

In the spirit of rugged individualism and self-reliance, some homeowners go further than doing some of the work themselves and rush right into acting as their own general contractor. Although most contractors, architects, and designers react with something close to horror at this prospect, it can be done successfully with the proper dedication and education.

One way to function as the general contractor for your renovation project is to hire each subcontractor under a separate contract, and take sole responsibility for coordinating and supervising their work. According to some experts, this can save you up to 30 percent of the cost of construction. But will it be worth it?

None of us would dream of setting a broken leg without medical training, but many of us think we can remodel a kitchen or build an addition with no trouble at all. And if *we* can't, then our brother-in-law or our cousin Freddie, neither of whom has any training either, can surely do it—and save us money. Why hire an expensive professional, we figure, when we can easily do it ourselves?

Underlying this attitude may be a certain snobbery about people who earn a living with their hands, wear jeans and T-shirts to work, and may not have a college degree. We fail, in this thinking, to take into account the years of experience that go into being a good carpenter or contractor.

Stephen M. Pollan, a financial consultant and coauthor with Mark Levine of several books on the economic aspects of home improvement, takes a dim view of the do-it-yourself impulse. In *The Big Fix-Up*

(New York: Simon & Schuster, 1992) Pollan claims that entire industries have sprung up to encourage the misapprehension that we can do as good a job as a trained professional. "The do-it-yourself industry will do anything to convince a customer they can actually do a good job," says Pollan. "And despite what the how-to books, genial television hosts, or the reassuring clerks at the home center say, you won't be able to do as good a job on your own. No book can give you the expertise and experience an electrician has acquired in 30 years of working on wiring. And no amount of well-intended advice can give you the knowledge necessary to coordinate, schedule, and manage a construction job."

We may think that managing a construction job is simple. And we may want to take on that role, eliminate the middle man, hire the subs ourselves, and save a bundle. But we fail to consider all the aspects of a renovation. In fact, it's really not so simple. Take a straightforward bathroom remodel, for example—just hire a plumber and a tile installer and it's finished in a few days, right? Well, not exactly. Whom do you schedule first—the tile layer or the plumber? Have you arranged for someone to remove the old fixtures and tile? Who will repair holes in the wall that this might create? What if you learn that you need a new subfloor? Who will clean up the mess and dispose of the debris?

Determined to prove we can do it ourselves, we also fail to consider the value of our time. As Stephen Pollan observes, "If you earn 40 dollars an hour, you really *save* money by paying a contractor 30 dollars an hour to do renovation work for you." And time is more than just money, suggests Pollan. Leisure hours are precious. Why not use them in relaxing and productive ways instead of doing poorly what a trained professional can do well?

A CAUTIONARY TALE: JESSICA'S STORY

Two people who would endorse Pollan's ideas are Jessica Tolliver Shaw and her husband, Nate. Seduced by the belief that they would save money and experience the hands-on satisfaction of working side by side on their own home, the couple decided to act as general contractors for the renovation of their 19th-century brick townhouse in Brooklyn. Adding considerably to these romantic do-it-yourself notions was the fact that the couple bought the house just months before their wedding.

Their story is an all-too typical tale about doing a potentially right thing for all the wrong reasons. First, they allowed themselves to be influenced by a family friend who had renovated several Brooklyn brownstones of his own and assured them that they could do it. Second, they failed to educate themselves about the scope of the job or to get any professional advice. "We thought it would be a simple project," says Jessica. "Update the kitchen, enlarge a bath, and refinish the floors. And we didn't think we needed an architect or general contractor to do that. We failed to see the big picture." Actually, they didn't even see the little picture very clearly. After they bought the house, they discovered many flaws they had not noticed before: faulty windows, mismatched moldings, crumbling plaster.

A third mistake was hiring workers without checking them out. According to plan, Nate and Jessica did all the demolition and hauled away the debris; then they called in an electrician who had been recommended by the family friend. After a couple of days the electrician asked who was handling various aspects of the project. "We didn't know," says Jessica, "because we hadn't thought things out very well. The guy said, 'My cousin does drywall,' and we said OK. And little by little the electrician sort of took over the role of general contractor."

Things started to go wrong when Jessica and Nate had trouble communicating with the electrician turned unofficial general contractor. "He talked down to us," says Jessica. "At first we assumed it was because he knew so much more than we did. But then we began to notice mistakes, such as poorly installed moldings. When we asked about it, he became defensive. We couldn't get a whole sentence out before he began interrupting and raising his voice. At that point, the only way to communicate was to raise our voices even louder, which we didn't always feel like doing. So we overlooked some things. In fact, we're still living with the clamshell molding that he installed in some of the rooms. It's not what we requested, and it's completely wrong for the house, but we didn't want to confront him about it."

Jessica recalls that she and Nate were so ill-informed that they could easily be talked out of things. For example, they wanted to save the original plaster walls if possible. But the family friend and the electrician persuaded them that the walls were beyond repair. When

Jessica suggested using a skim-coat over them, they claimed never to have heard of such a thing. And down came the walls, to be replaced by drywall. "The fact that we lost control of the walls is my biggest regret," says Jessica. "Even less-than-perfect plaster would have been OK. We would have loved the bumps and irregularities."

Living in a state of semi-denial, hoping for the best, and reluctant to confront the electrician, who by now had become almost impossible to talk to, Jessica and Nate went along with the program for a while longer. "Sometimes, we didn't have the energy to correct him," they recall. Then a cabinet installer came in to lay out the kitchen; he pointed out so many electrical wiring—and other—mistakes that the that the couple were forced to face facts. Says Jessica, "Holes were cut too big for the switch plates and outlets, moldings weren't mitered and didn't lay flat, doors didn't hang right, light fixtures weren't centered, just put up randomly."

Finally, they fired the makeshift contractor and hired a bona fide general contractor to finish up the kitchen and correct the wiring. "We spent a lot to fix things," says Jessica. "Plus we were planning a wedding at the same time. It was a very stressful time." They're lucky they found someone to come in and do damage control; many remodeling professionals are reluctant to take on projects that have gone bad, figuring they'll be too difficult and time-consuming to fix.

Jessica's advice to anyone who wants to be her own g.c.: "Don't do it. Not unless you're willing to spend all of your time at it, far longer than you think you will. We would never do it again. We'd hire an architect or designer to oversee everything and we'd be involved only to approve of things."

Here's the good news: Nate and Jessica learned their lesson. "We have one room left to work on," they say. "And in that room we're doing *nothing* ourselves!" In addition, after correcting their initial mistakes, they finally got the house they wanted. And best of all, they're still married.

TWO SUCCESS STORIES

As we know, most architects and contractors shudder at the very thought of homeowners acting as their own general contractors. But Laura Glassman didn't let that stop her. She and her husband Len built their dream house in Delhi, New York, a couple of years ago, with Laura acting as general contractor. Which is one reason, she thinks, that their dream house actually became a reality.

Glassman had some experience. She had built a house some years before with a carpenter friend and had supervised the renovation of her husband's ophthalmology office in New Jersey. In addition, when she and Len began planning their own house, she signed up for several classes at Heartwood School in Washington, Massachusetts, to further educate herself, studying such subjects as building, acting as your own contractor, and designing your home yourself.

Glassman also had the time. She was able to devote herself 100 percent to the project and oversee every aspect of it. "It took about 18 months full-time for a 4,600-square-foot house," she estimates. Even for a smaller house it would have consumed the better part of a year, she thinks.

The project was hands-on from the very beginning. Because Glassman knew precisely what she wanted, she interviewed architects until she found one who allowed her to be part of the design process. Then, plans in hand, she hired a local general contractor who was willing to function on this job strictly as the builder. "He was a little pessimistic about my abilities, I think, but finally he said, 'OK, let's give it a shot.' "

Why take on all this work and potential headaches? "To guarantee the result I wanted and to save money," says Glassman. She cut costs, she claims, by eliminating the middleman and hiring all the subs herself. "They were bidding expressly for me, and I believe I got the best price," she says. The contractor built the house up to the primed drywall, then Glassman bid out all the other work—plumbing, wiring, tile installation, painting, cabinetmaking. "I got the best people I could find," she says, "sometimes even bringing them in from out of town, and they all dealt directly with me."

Glassman's story has a happy ending. She and her husband got the house they wanted, and she loved supervising the creation of that house. It worked because she had time, familiarity with the process, and a burning desire to have the experience.

Carla McClinton, another successful do-it-yourselfer, acted as her own general contractor because it was the only way she could afford to build her 6,000-square-foot Clinton, Maryland, house, which she designed herself. And to make it possible, McClinton did essentially the same things Laura Glassman did. She educated herself, taking courses on construction at her local community college, reading books on building your own home, and watching new houses go up. "I went onto residential construction sites, hung out with the guys, and asked questions," she says. "Usually, people were very helpful. I got a lot of valuable information that way. I made sure I knew the whole process ahead of time."

Can You Pass the DIY General Contractor Test?

The temptation to act as your own g.c. and save perhaps 30 percent of your project's cost is a strong one. But look before you leap, urges Smart Consumer Services' N'ann Harp, a former general contractor. "It's not for the faint of heart," says Harp. She suggests that anyone considering this bold move ask themselves a couple of soul-searching questions first: (1) Have I thoroughly investigated every step of the process—financing, design, site work, permits, materials estimating and ordering, crew scheduling, contracts, and final inspection? (2) How about management skills? Do I like working with people; do I have the ability to direct them; do I have the confidence to make firm decisions?

Anyone who can respond with a yes to these questions will probably find it rewarding and fun to act as his or her own contractor, says Harp. And, she adds, "After you've taken on a major project like this, you come away with increased regard and respect for contractors."

If you're still not sure, take a look at a couple of do-it-yourself television shows or videos, which you can probably easily get from your local video-rental store or library. If one picture is, in fact, worth a thousand words, watching people engaged in work that's similar to what you want to undertake could very well help you make your decision.

Like Glassman, McClinton was prepared to dedicate a great deal of time to the project, although she had to do some fancy footwork to also put in a full day at her job as a realtor. "I went to the site morning, noon, and night literally," she says. "Before work, when I gave the subs their daily schedule; on my lunch hour; and after work. And during the day, my 23-year-old son went by regularly to check on things."

Financing, not an issue for Len and Laura Glassman, was a big challenge for McClinton. "Three banks turned me down," she says. "They wanted me to have a professional builder, and they didn't believe my cost estimates were realistic. One bank practically accused me of fraud. I finally got the loan from a bank that does a lot of construction financing; they saw what I was trying to do and they believed I could do it." So successful was McClinton that both she and her house were featured on a national cable TV show.

two

Budgeting and Finding Contractors

You made a smart choice when you decided to remodel your house, and here are some of the reasons, as suggested by the Remodelors Council of the National Home Builders Association:

- Remodeling allows you to tailor your house to your specific needs. The only other way to do this is to design and build a brand-new custom home, and that will cost you a great deal more money.
- Remodeling, instead of moving, allows you to hold on to important and treasured parts of your life such as a familiar, well-loved neighborhood, good schools, old friends.
- Remodeling makes efficient use of your financial resources. According to the American Homeowner Foundation, selling your home and moving typically costs about 8 to 10 percent of the value of your current home. And much of this percentage goes for moving expenses, closing costs, and broker commissions, expenditures that

have no impact on the quality of your new house.
- Yes, remodeling is stressful, but nowhere near as stressful as moving.

ASSESSING THE PROJECT YOU'RE PLANNING

But before you embark on this worthwhile endeavor, be sure the project you are planning makes good sense for your needs, your living situation, and your budget. Careful planning at the outset will help you organize your project and determine a working budget.

For instance, say an updated kitchen is your goal. If your existing one is more than 15 or 20 years old, there probably is a lot wrong with it—old appliances, insufficient food-prep space, not enough storage, worn countertops and floors. But before you leap into action with plans for a new kitchen, think about the big picture. What is the overall house like? Is it in generally good shape inside and out? Will sprucing up the kitchen put the other rooms to shame? Is the kitchen the only area that needs improvement? If not, could you solve several problems at once by expanding the remodeling project a little—adding a bath on the first floor or opening the kitchen into the living room or family room, for example? Both of these changes would almost certainly improve the resale value of your house. If your neighborhood is solid, your house in good shape, and you plan to sell within the next few years, it makes sense to put some money into a really nice kitchen that will become an asset.

But if you can't afford whatever it takes in your locale to finance a professionally designed kitchen remodel, you might choose to touch up the kitchen yourself, freshening old cabinets with paint or varnish, laying a vinyl floor, installing a new range or some laminate countertops. Because many home buyers end up remodeling kitchens to suit themselves, you might even be better off doing nothing and selling the house as is.

Whether you do it yourself or hire a professional, install the best quality fixtures, appliances, and hardware you can afford. Prospective home buyers have become increasingly sophisticated and knowledgeable about products over the last few years, and they will respond positively to any well-known name-brand, even if it is at the low end of the manufacturer's line. Prospective buyers by and large will prefer a small house with high-quality features and excellent workmanship to a large one that's been poorly renovated.

Beware of over-improving. Real estate professionals often cite the 30-percent formula. As a general rule, they say, any home improvement increases the value of a house by only 30 percent over the cheapest house in that neighborhood. Gear your project to move you from the low end of your neighborhood's price range to the high end. If you go beyond, you're not going to recover your remodeling costs.

If you're planning to sell soon, now is the time to set up an accordion file or other appropriate, efficient holder of receipts for all the bills you'll be paying during your project and mark it "Save for resale." Nonexistent or incomplete remodeling receipts will undermine the strength of your seller's position and the basis of your asking price. In order to recoup your investment, you must be able to prove what you spent. (You'll also need these records for tax purposes.)

Perhaps you don't care about resale value. Maybe you plan to stay in this house forever. In which case you would make changes only for yourselves, not with an eye to your home's future salability.

The National Association of the Remodeling Industry (NARI) sums it up this way: Spend as much money as necessary to create your dream home if you are planning to live in the house for five more years, and if you can afford to do so. However, if you are planning to move, remodel within the standards for houses in your neighborhood and within a reasonable budget.

But if you do want an informed opinion about how your remodeling project will affect the resale value of your house, contact a local house appraiser or a real estate broker, preferably someone who has been in your area for some time. For a small fee—$150 or $300, for the appraiser; less, or perhaps no fee at all, for the real estate person—he or she can give you a good idea whether your proposed renovation will increase the value of your property and the salability of your house.

Some remodeling projects offer reliable payback benefits, that is, they increase the value of your house so handsomely that you recover 80, 90, even 100 percent of their cost at resale time. A neutral-color exterior paint job, attractive landscaping, and conventional kitchen and bathroom renovations (ones that don't add pricey finishes or appliances or unusual paint colors) are generally considered by real estate and remodeling professionals to have high return value. If you add a swimming pool, tennis court, outdoor hot tub, or any ultra-expensive interior improvement, such as custom cabinets or luxurious carpets, you may not recover the cost at resale time. However, if one of these projects will improve the quality of your life and you don't mind not recovering its cost, by all means go ahead.

WHAT'S THE PAYOFF?

Remodeling magazine publishes a yearly "Cost vs. Value" report that contrasts the cost of home improvements with their recouped value when the house is sold within one year of project completion. Here, from 1999's study, are some projects with profitable payoffs. Don't be put off by the fact that there are no 100-percent recoveries of remodeling money here. First of all, that happens only in very hot real-estate markets. Second, the homeowners have had full use of the improvements for up to a year, which adds to their value. Third, any home-improvement project that offers a 70- to 80-percent return is an excellent investment.

PROJECT	JOB COST	RESALE VALUE	PERCENT RETURN
Minor kitchen remodel	$8,655	$7,041	81%
Project description: In a functional but dated 200-square-foot kitchen with 30 lineal feet of cabinetry and countertops, refinish existing cabinets, and install new energy-efficient wall oven and cooktop, laminate countertops, midpriced sink and faucet, wallcovering, and resilient flooring. Repaint. Job includes new raised-panel wood doors on cabinets.			
Bathroom addition	$13,918	$10,000	72%
Project description: Add a second full bath to a house with one or $1\frac{1}{2}$ baths. The 6-by-8-foot bath should be within the existing floor plan in an inconspicuous spot convenient to the bedrooms. Include cultured marble vanity top, molded sink, standard bathtub with shower, low-profile toilet, lighting, mirrored medicine cabinet, linen storage, vinyl wallpaper, ceramic tile floor, and ceramic tile walls in tub area.			
Bathroom remodel	$9,135	$6,442	71%
Project description: Update an existing 5-by-9-foot bathroom that is at least 25 years old with a new standard-sized tub, toilet, and solid-surface vanity counter with integral double sink. Install new lighting, faucets, mirrored medicine cabinet, ceramic tile floor, and ceramic tile walls in tub/shower area (vinyl wallpaper elsewhere).			
Family room addition	$30,960	$21,868	71%
Project description: In a style and location appropriate to the existing house, add a 16-by-25-foot room on a new crawl space foundation with wood joist floor framing, wood siding on exterior walls, and fiberglass shingle roof. Include drywall interior with batt insulation, tongue-and-groove hardwood floor, and 180 square feet of glazing (including windows, atrium-style exterior doors, and two operable skylights). Tie into existing heating and cooling.			

Major kitchen remodel $31,090 $21,888 70%

Project description: Update an outmoded 200-square-foot kitchen with design and installation of a functional layout of new cabinets, laminate countertops, midpriced sink and faucet, energy-efficient wall oven, cooktop, ventilation system, built-in microwave, dishwasher, garbage disposer, and custom lighting. Add new resilient floor. Finish with painted walls, trim, and ceiling. Include 30 lineal feet of semicustom-grade wood cabinets and counter space, including a 3-by-5-foot center island.

Master suite $42,826 $29,134 68%

Project description: On a house with two or three bedrooms, add a 24-by-16-foot master bedroom suite over a crawl space. Bedroom will include a walk-in closet. Master bath will include dressing area, whirlpool tub, separate ceramic tile shower, and double-bowl vanity. Bedroom floor will be carpeted; floor in bath will be ceramic tile.

Attic bedroom $28,654 $18,753 65%

Project description: In a house with two or three bedrooms, convert unfinished space in attic to a 15-by-15-foot bedroom and a 5-by-7-foot shower/bath. Add a 15-foot shed dormer and four new windows. Insulate and finish ceiling and walls. Carpet unfinished floor. Extend existing heating and central air conditioning to new space. Retain existing stairs.

Two-story addition $73,553 $45,910 62%

Project description: Add a 24-by-16-foot two-story wing, over a crawl space, with a first-floor family room and a second-floor bedroom with full bath. Include a prefabricated fireplace in the family room, 11 windows, and an atrium-style exterior door. Floors are carpeted, and walls are of painted drywall. The 5-by-8-foot bathroom has a fiberglass bath/shower, standard-grade toilet, wood vanity with ceramic tile sinktop, ceramic tile flooring, and mirrored medicine cabinet with light strip above; bathroom walls are wallpapered. Add new heating and cooling system to handle addition.

Siding replacement $5,838 $3,487 60%

Project description: Replace 1,250 square feet of existing siding with new vinyl siding, including trim.

Window replacement $7,531 $4,226 56%

Project description: Replace 10 existing 3-by-5-foot windows with vinyl-clad windows, including new trim. Replace sash, frames, and casings.

Deck addition $8,022 $4,346 54%

Project description: Add a 16-by-20-foot deck of pressure-treated pine supported by 4x4 posts set into concrete footings. Include a built-in bench, railings, and planter, also of pressure-treated pine.

Home office $8,356 $4,219 50%

Project description: Convert an existing 12-by-12-foot room to a home office. Install custom cabinets to include desk area, computer workstation, overhead storage, and 20 feet of laminate desktop. Rewire room for computer, fax machine, and other electronic equipment, as well as cable and telephone lines. Include drywall interior and commercial-grade level-loop carpeting.

Reasons to Renovate

- You want more living space (but don't want to move, can't afford to move, or haven't been able to find another house that suits your needs).
- You want to update the functionality of your existing house.
- You want to increase the pleasure you get from your house.
- You want to add character to a plain-Jane house.

EDUCATE YOURSELF BEFORE YOU BEGIN

Whatever your reason for renovating, you're bound to have some anxieties—confusion about the process, dread at the thought of going through the mess and disruption, apprehension about costs, fear of losing control of the project, and concern about the money. Therefore it's a good idea to know what you're in for before you embark on your project. A little homework beforehand will not only prepare you for the enormity of the undertaking but also teach you how a renovation happens, how it will change your life, what part you will play in it, and, very roughly, what it could cost. And because forearmed is forewarned, a little knowledge can protect you against a contractor who might try to persuade you to contract for more work—and more money—than is necessary.

You don't need to know everything, of course. Just enough to make you an educated consumer.

Laura Glassman, the woman who acted as her own contractor (see Chapter 1), educated herself about her project before she did anything else, which is one reason that it was such a success. If you have only a dim idea of what you want, you'll have trouble communicating with the contractor, the job will probably not go smoothly, and you could even end up with a renovation that doesn't suit your needs and makes you unhappy.

Glassman knew all about the house she wanted to build. She knew its size; its architectural style; the materials needed inside and outside; the number of rooms and their functions, sizes, and positions in the floor plan. Before she placed the rooms in the plan she first studied the way she and Len, their friends, and their family would use the space; how the sun would travel through the house; what views the windows would frame and how those view would change with the seasons. She also did her homework on appliances, cabinets, bathroom fixtures and fittings, doors, windows, lighting fixtures, switch plates, cabinet hardware, in short, everything.

One way to arrive at precision like this is to work with an architect or designer, of course, but even if you hire a design professional, you should approach the initial interview with a clear idea of what you want. You don't need to know all these details right away; but before you hire a contractor, you should have a clear idea of what you want done. Not a vague statement like "I want to add a master bedroom and bath to my house somewhere." But something definite: "I want to add a one-story addition onto the back of my house to hold a 450-square-foot master suite with high ceilings and nice views of the garden. And I want the design to be in keeping with the existing house."

If you know what you want but are unclear about how to get it, you can use preliminary meetings with a design professional to lend focus to your project. However, before you commit yourself to any course of action, be certain that the pro's suggestions truly do meet your needs and desires.

It's also a good idea to know what to expect in terms of physical disruption. We go into detail about this in Chapter 5, so at this stage all you need is a general picture. Will there be extensive demolition, excavating for a new foundation in the middle of your rose garden, debris and mess piling up, then being noisily hauled away? Will cement trucks arrive, making more noise and mess? Will load-bearing walls have to be shorn up, floors torn up and redone? Will you be without water or electricity for a period of time?

How-to videos and television shows such as *This Old House* and *Hometime* can help you visualize what the job will be like. Even if you don't intend to do any of the work yourself, just watching other people do home-improvement projects like the one you're contemplating will give you a vivid picture of what you're in for.

Knowing the overall scope of the project and being familiar with the materials, finishes, appliances, cabinets, fixtures that you like best (and are in keeping with your budget) will save you time and money. If you have done your homework before you start, you will not go into shock to learn that the granite counters you dreamed of for your kitchen could easily run you as much as $400 per linear foot installed. Maybe you'll go for the granite in spite of the cost; maybe you'll substitute a man-made solid-surfacing material that looks almost as good but costs less. But thanks to your preplanning, the remodeling project won't have to come to a screeching halt while you get over your shock and find a substitute. The same thing might happen with windows. The architecturally arresting grouping of unusually shaped windows you envision for one wall of your proposed addition could set you back a small fortune if you select custom units. A similar grouping of stock windows could be just as effective and would cost a lot less.

Picture Your Project

One of the simplest and most useful tools you can use to assess your project is a clipping file. And later it can help you communicate with your contractor, designer, or architect. Get started early, several months before you are ready to begin your project or commence your search for a contractor. Flag or tear out photographs in books or magazines that visually describe an architectural detail, layout feature, decorative treatment, or surface (paint, wallpaper, paneling, flooring) that appeals to you or solves a problem that exists in your present house. You can also make a sketch of how you want the finished project to look. Don't let lack of drawing talent stop you; even a very rough sketch of what you want could be a useful tool. A file such as this will clarify your needs and help you present them effectively to the contractors you interview. The file will also serve as an accurate indicator of color and other features, no matter how skilled you may be at verbal communication. For professionals in the highly visual field of home improvement (architects, designers, contractors, and the like) pictures are worth more than words. All of them will appreciate your efforts to be as precise as possible with a clipping file.

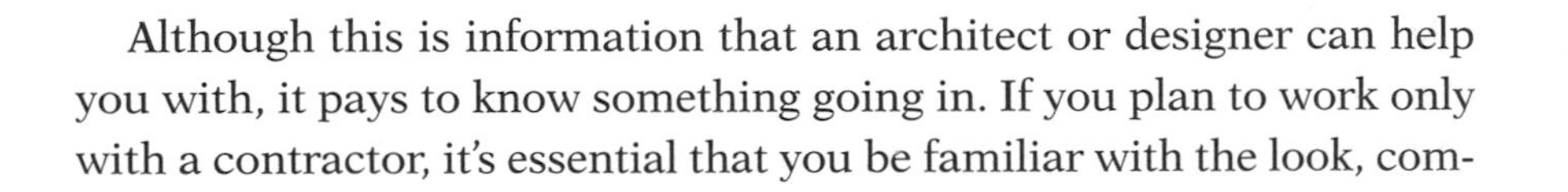

Although this is information that an architect or designer can help you with, it pays to know something going in. If you plan to work only with a contractor, it's essential that you be familiar with the look, composition, and general cost of the materials, finishes, and fittings that your project will require. Don't leave those choices to your contractor. He or she is an expert on building, not a designer. Know what you are asking people to create for you. Otherwise, the creation will be something that your architect, designer, or contractor wanted and not the renovation that meets your practical and aesthetic needs.

Part of the self-education process is knowing what the products and materials you select will look like once they're installed. Don't make the mistake that Margaret Harris did. When she updated her 19th-century house in a small Indiana town, she allowed her contractor to talk her into vinyl-clad windows as replacements for the worn and leaky original wood units. She approved the order knowing the replacements would be durable and economical but not having any clear idea of what they would look like. A good part of the job had been done before she realized how inappropriate and out of place they looked on the facade of her fine old house. She put a halt to the work, paid for what had been done, and eventually chose windows that looked right and were not significantly more expensive. But it was a costly mistake.

Once you've come up with products and materials you think you would like—or that someone has suggested to you—try to see them in the flesh. Visit a showroom or attend home shows to see what a Corian countertop looks like; go to a friend's or neighbor's house to view replacement windows; stop by houses that have been recently sided with the vinyl that your contractor is touting as dependable and economical. In addition, many manufacturers will send you samples of products you're interested in.

Pat McKernan, who remodeled her Tucson, Arizona, kitchen, had to struggle to stay within her strict budget. Her contractor, trying to be helpful, suggested a new product for the floor, a laminate designed to look like wood. Magazine photographs of the product looked good, as did the samples in McKernan's local flooring store. But she knew an eight-inch-wide, three-foot-long sample wouldn't tell the whole story. She asked to see a real installation and the salesman sent her to an office down the street where the product had just been installed. One look told her that it wasn't for her. That little trip down the street saved her from an expensive mistake.

Do a little comparison shopping for specific products at home centers, appliance stores, showrooms of various kinds. You may amend your wish list and your budget as you get a realistic picture of what materials and products actually cost and what they look like.

Ideally, your pre-renovation homework will include giving yourself plenty of time to search out and select a good contractor. This process requires patience, research, and attention to detail and could conceivably take a couple of months. If you don't build in enough time for it, you're apt to make a hasty choice that could cause you a great deal of trouble down the line.

Heads Up—Learn the Language

As you do some simple research on how a house is constructed and what's involved in a major remodeling project, you will begin to get familiar with some of the terms that builders, contractors, plumbers, and electricians use. (You can also study the Glossary at the end of this book.) Familiarity with these terms will be a great asset. When people are using words we don't understand, we tend to feel uninformed and uncomfortable; perhaps we even go so far as to pretend we know what they're talking about when they toss around such terms as "header," "mullion," and "fascia." And in this muddled state, we may agree to things we don't understand instead of asking questions that we think may make us look as stupid as we feel. Once you know what these esoteric terms mean, no one will be able to talk you into something you won't really like or can't afford.

MAKE FRIENDS WITH YOUR BUILDING DEPARTMENT

Before you get too deep into your remodeling dreams, you'd better find out if they can come true. Now, before you have spent any money, take a trip to your local building department. For a small project, a phone call is probably sufficient. Although laws vary from state to state and from one town to another, you'll probably need a building permit if your project changes the function of a room; or if it includes the addition or alteration of structural, electrical, plumbing, or ventilation work; or if it involves the number or size of windows. But you're not there to apply for a building permit; your contractor will do that at a later date. The purpose of this trip is to unearth any potential problems posed by your project.

Every community has rules and regulations about what can be built and what can't. Take setbacks, for instance. The restrictions in your neighborhood may prevent you from building the addition you had your heart set on. Or the luxurious master bath you wanted to add, complete with hot tub and steam room, may require an expensive septic-system upgrade that will put you way over budget.

In addition, your project may require zoning approval. Local jurisdictions have the right to limit the height of your house and impose setback ordinances that regulate the distance from the lot line to the point where improvements may be built. You may have a neighbor's approval to build close to the property line that you share, but that's not enough; you must also get approval from the zoning board.

If you live on or near a body of water or a flood plain, or if your property abuts conservation land, you should meet with your local conservation commission or environmental protection agency. If your house or your neighborhood are old and historic, check with the historical society before you make any definite plans. These organizations often wield a great deal of local power and may have a lot to say about the project you are planning. It pays to work with, rather than against, them.

And if you're thinking about circumventing the permit process, chafing against the idea of red tape and government interference, think again. It's embarrassing, inconvenient, and expensive to be found out by the building department. At the very least you will have to get the permit then and pay some fines and perhaps rebuild to code. In a worst-case scenario, you could be forced to stop work and wait for a long period of time, with your house torn apart, before you are allowed to resume, or be forced to tear down what has already been built. And even if you get away with it now, you'll have to obtain the proper permits eventually in order to sell your house, which could be a whole different kind of headache. (For more information on building codes and permits, see listings in Chapter 7.)

DEVELOP A BUDGET

Homeowners on the verge of a remodeling should also know roughly how much the job is likely to cost and how much they are willing to spend. Do not begin a renovation project without knowing what your costs are likely to be. If you develop a working budget at the beginning, you will be a great deal less likely to waste time, effort, and money pursuing a makeover that you can't really afford. By the time you reach the bidding stage, you will have a reasonably clear picture of what the remodel will cost.

"People really don't have a clue about what remodelings cost," says Kenneth Skowronski of the design-build firm KS Remodelers. "We design what people tell us they want, price it, and then present a proposal to the clients. This is the moment of truth, the reality check. Often the clients are totally shocked, and sometimes they even get angry. They are reluctant to spend what is necessary to get what they want.

"It's important to do your homework before you call a contractor," says Skowronski. "Compare the job you're planning with similar ones you've seen, so you can get at least a rough sense of what your project might cost."

Try to avoid the "what's-it-going-to-cost" impasse that N'ann Harp of Smart Consumers Services describes. The scenario goes like this. Homeowners open with a vague statement about their proposed remodeling: "We'd like to enlarge our kitchen and maybe knock out a wall and add a banquette; we also need a pantry and a second sink."

"No problem," says the contractor. "How much do you have to spend?"

The question is logical and legitimate. But most homeowners, who haven't an inkling what things cost, panic. "Oh-oh, here it comes," they think. "We're being ripped off already. We only want to know how much it's going to cost. We don't want to tell anyone how much money we've got!"

Now on the defensive, the homeowners are aware that they don't know enough about kitchen remodeling to make an informed decision about how much to spend. But they don't believe for a minute that this stranger sitting at their dining table is going to give them any straight answers. In fact, they probably think he or she is attempting to take advantage.

Right about now the contractor may be thinking something like this: "Oh, great. Another couple who hasn't done any homework. I'm going to end up doing 10 estimates in decreasing dollar amounts because these people want me to guess how much they can spend on the project. I hate working this way."

What typically occurs next, says Harp, is a slow torturous dance in which the contractor delicately unearths the financial information he or she needs to give the homeowners good recommendations and information that will help them make decisions. For their part, the poker-faced homeowners are doing everything they can to avoid giving clues about the status of their budget for fear of being victimized.

There is no surer way to get off on the wrong foot with a contractor, Harp claims. "The unspoken, underlying feelings of anger and distrust created during these early meetings are the most insidious, destructive, and common features of poor homeowner-contractor relations. They set you up for more severe problems down the road when the job is under way and some minor miscommunication suddenly mushrooms into a toxic cloud of all your suppressed anger and distrust. This may sound like an exaggeration but it is not," Harp insists. "A little stress is all it takes to bring these unpleasant ghosts back to haunt you."

The negative effects of this early suspicion and distrust are almost impossible to overcome later in the relationship. But the problem can be avoided if you will prepare yourself with a rough idea of what your type of project might cost before you meet with the first prospective contractor. And even before that conversation takes place, be sure you know how much you can realistically afford to spend.

Don't force the contractor to play "What's the budget?" with you, Harp urges. Prepare yourself properly so that you can tell him or her at the outset what you have budgeted for the renovation. The contractor will then become your ally rather than an imaginary adversary and will probably start sharing inside tips for saving money or solving your particular challenges.

One way to find out what you can afford is to visit a lender to prequalify for your remodeling funds. The lender will request information on your employment, income, assets, debts, and credit history, and then calculate the maximum loan you could obtain. Keep in mind, however, that the loan is not guaranteed at this point. If the information you give at prequalification doesn't check out, or if it changes, when you submit your final application, you may not get the loan.

There are a couple of ways to come up with likely costs. One is to find out the average cost per square foot of construction in your area and then multiply that by the approximate square footage of the addition you are building or the rooms you are renovating. Double this figure to estimate what finish carpentry might cost. Then, to this figure you will have to add the approximate costs of surfaces (flooring, counters), equipment (appliances, cabinets, lighting fixtures), and finishes such as paint or wallpaper. This will be a very rough figure, but it is a good start. To ascertain the average cost per square foot for construction in your area, call such local experts as real estate agents, house appraisers, or the local building authority, or contact the nearest office of the National Association of Home Builders (NAHB) or the American Institute of Architects (AIA) (See Chapter 7 for details.)

Another way to figure out how much your project will cost is to have a contractor come in early on, look over what you want to do, and give you a ballpark figure. Make clear to the contractor that you have not yet reached the hiring stage and are not asking him or her to bid on your job, but that you *are* willing to pay a fee for a rough professional assessment of what the job might cost. (The fee would probably be deductible if you end up hiring the same contractor to do the job.) This one-time service can be invaluable in helping you realistically assess your project and plan its execution. Not every contractor has the time or inclination to offer this kind of service, so you may need to look around to find one who does.

Cost-estimator guides are also very helpful tools. Geared toward contractors performing work with their own crews, these user-friendly handy guides estimate the costs to the professional of hundreds of construction, repair, and remodeling jobs: demolition, excavation, wiring, plumbing, heating and cooling, rough and finish carpentry, door and window installation, and more. The estimates are based on information from contractors, engineers, design professionals, construction estimators, materials suppliers, and manufacturers. The guides are updated frequently to keep the information current, and they also help you adjust average national costs up or down to coincide as closely as possible with going rates in your area.

The granddaddy of these publications is the *Home Tech Remodeling and Renovation Cost Estimator* published by Home Tech in Bethesda, Maryland. Some other guides include the *National Construction Estimator* (Craftsman, Carlsbad, California), the *Remodeling Costbook* (BNi Building News, Los Angeles), and *The Means Building Construction Cost Data* (Robert S. Means Co., Duxbury, Massachuetts).

The estimators are somewhat pricey—from about $50 up to more than $100—but probably worth it if you're planning a large remodeling project. And of course or you can always look for one at your local library.

Another way to estimate costs is to talk to friends, neighbors, and associates who have had similar work done. People generally love to talk about their remodeling projects. Some are unwilling to talk about money. But if you explain why you want to know, they may tell you at least approximately what their project cost, especially if they are happy with the results.

Once you have established a budget, stretch it a little—by a couple of thousand dollars if you can—to prepare for the unexpected things that always happen in a remodeling.

10 Ways to Stay on Budget

- Plan cautiously. Make all the changes you want on paper; they're expensive later on.
- Prioritize. Decide where to economize and where to focus your funds.
- Shop critically. Avoid one-stop buying; you may end up paying too much for the convenience.
- Stick to standard and stock choices. Find out how much special finishes or colors will add to your costs.
- Understand the differences in materials. Consider long-term value as well as initial cost.
- Don't be swayed by status. Does that stylish product really suit your needs? And will you still like it next year?
- Refurbish and recycle. Can you reuse windows, doors, appliances, and other equipment instead of replacing them?
- Keep the structural framework. Before adding on, explore the more economical possibility of reconfiguring the existing space.
- Pay for professional advice: A skilled designer or architect can help stretch your budget.
- Do some of the work yourself but take care not to overestimate your zeal or skill.

DON'T GET IN OVER YOUR HEAD

When he remodeled his house in Sneden's Landing, New York, Gary Bennett failed to do his homework or develop a budget. And he paid a high price for it, eventually having to sell the place to try to recover some of his money. Several years ago, Bennett, a film maker and teacher of film acting, and his wife, Alyssa, fell in love with a charming historic house in this little village on the Hudson River. Although Alyssa's brother, an architect, tried to discourage them from buying it, warning them that it needed a lot of updating, they were too smitten to listen.

Having bought the place at what they thought was a bargain price—and it was a bargain compared to the costs of surrounding houses—the Bennetts set about making it livable. "We loved it so much," says Bennett, "that we decided we would do whatever it took. We pictured ourselves living here forever, raising our children here, dying here."

With Alyssa pregnant with their first child, charming idiosyncrasies such as the tiny rooms (too small for most of their furniture) and low doorways began to seem less charming. So did the fact that the only built-in heat was the fireplace; the other rooms relied on space heaters. Instead of pacing themselves, the Bennetts decided to do all the work right away. They installed a heating system and enlarged the 600-square-foot house with a 2,000-square-foot addition.

With no clear idea of what the total project might cost them or what they were spending week to week, the Bennetts were headed for trouble. "Once the addition was framed and roofed, I thought we had spent the bulk of the money," says Gary Bennett. "That was a big flaw in my thinking; I believed that the inside stuff—drywall, flooring, doors, windows, cabinets, and other finish work—would be less costly than the structure." Actually, just the opposite is true, as he eventually discovered.

But before the day of reckoning came, the Bennetts continued to make choices. "My wife liked tumbled marble for the kitchen floor. If she had known how much it cost, she would have chosen something else. But I wanted so much to please her that I didn't mention how expensive it was, and I really didn't grasp the cost myself. Someone said $40 a square foot, but that didn't sound so bad. I never calculated delivery, installation, and all the rest. It ended up costing $9,000.

Calculate Your Budget

Complete the following worksheet to determine how much you can afford to spend on a remodeling project.

Current assets

Cash on hand . . . . $__________
Savings accounts . . . . $__________
Cash value of stocks . . . . . $__________
Mutual funds . . . . $__________
Bonds . . . . $__________
Life-insurance cash value . $__________
IRAs. . . . . $__________
401(k) plan . . . . $__________
Keogh plan . . . . $__________
Employee savings plan . . . $__________
Pensions . . . . $__________
Real estate . . . . $__________
Other. . . . . $__________
Total assets . . . . $__________

Current liabilities

Installment loans . . . . $__________
Credit-card balances. . . . . $__________
Student loans . . . . $__________
Other debts. . . . . $__________
Total liabilities . . . . $__________

Subtract your total liabilities from your total assets.
This is your net worth . . . . $__________
Emergency funds (six months' income is suggested) . . . . $__________
Subtract your emergency funds from your net worth. The remainder is the amount your have available for a renovation.
. . . . $__________

Current annual income

Gross salary . . . . $__________
Alimony . . . . $__________
Child support. . . . . $__________
Interest. . . . . $__________
Dividends. . . . . $__________
Tax refunds . . . . $__________
Other . . . . $__________
Total annual income. . . . . . $__________

Current annual expenses

Rent or mortgage payments $__________
Food . . . . $__________
Clothing. . . . . $__________
Transportation . . . . $__________
Medical/dental. . . . . $__________
Insurance premiums
Life . . . . $__________
Auto . . . . $__________
Home or renter's . . $__________
Other . . . . $__________
Tax payments. . . . . $__________
Utilities. . . . . $__________
Savings . . . . $__________
Tuition/day care . . . . $__________
Alimony/child support . . . . $__________
Loan/charge account
payments . . . . $__________
Recreation/entertainment . $__________
Other. . . . . $__________
Total annual expenses. . . . $__________

Subtract the total of your current annual expenses from your total current annual income. This figure is the amount you can spend per year on renovation . . . . $__________

Divide this amount by 12 to find the monthly loan payment amount that you can afford.
. . . . $__________

And once you've spent that much on your kitchen floor, you don't want to put in discount-store cabinets. Things added up fast, and the further we got into it the harder it was to stop. 'We'll pull out,' I thought. 'We'll be all right. After all, we're going to live here forever. We'll have plenty of time to recover.' " The Bennetts have recovered now, but they had to sell their house in order to do it. "Romantic notions got in the way of common sense and led us into trouble," says the rueful Bennett, "that, and not really being in control of what was happening."

Heads Up: Cut Corners, but Not Too Close

Remodeling costs a lot of money; count on it. You will be unpleasantly surprised when you receive estimates and bids from the professionals involved in your project; and you may even cringe at the price of the windows, doors, appliances, cabinets, carpeting, and so forth that you may have selected. Fortunately, there are ways to shave a little money off all these things, which you can discuss with the professionals involved in your project. But don't do anything drastic to save money, such as going for the contractor with the lowest bid even though you had some uneasy feelings about him or her, or deciding to do most of the work yourself although you lack the experience.

Trying to save money is what gets people into the most trouble with home-improvement projects. It is better to postpone the job until you can more easily afford it than to cut corners in crucial areas.

FINANCING THE PROJECT

At this stage of the process, it's a good idea to start shopping for financing, beginning with a visit to your local banker. He or she will be able to tell you how much money you qualify for and will discuss the financing options available. But don't stop there. Shop for the best interest rate at several other banks and other types of lending institutions, such as savings and loan associations, credit unions, and mortgage companies.

If your project is small, say, $3,000 or less, it might be wise to **pay cash** and avoid costly and time-consuming paperwork, finance charges, and interest. Cash financing might be smart for a larger project, too, provided that the interest income you lose by using your own money is more than offset by the interest rate a loan would cost you. For example, withdrawing some of your savings, for which you earn 4 percent interest, might be an attractive prospect compared to taking out a small loan at 10 percent interest. But devoting savings to a renovation will decrease your liquidity. Will you still have enough cash on hand to see you through an emergency?

Credit-card financing is convenient for small projects that cost about $3,000 or less and can be paid off quickly. Otherwise, the high interest rates make this method of financing too expensive.

For a project that falls between $3,000 and $10,000, the wisest choice is probably an **installment loan** of some kind. With a standard home-improvement loan, which is designed especially for remodelings and is secured by your home itself, payments are generally spread out over five years or so. On the plus side, these loans are usually easy to obtain and the interest is deductible. The downside: an abbreviated payoff period and high interest rates.

Another possibility is an **unsecured, or personal, loan,** based solely on your ability to pay. The bank will not specify what this money is to be used for, but it will want you to pay it off quickly, typically in one to three years, and the interest rate will be high.

For renovations of $10,000 or more, a **home equity loan** (sometimes referred to as a second mortgage) may be your best bet. Among the most popular sources for remodeling money, this type of loan allows you to take advantage of the equity you've built up in your home, which is the difference between its current market value and the balance of your mortgage. If, for example, your house appraises at $200,000 and you still have $100,000 to pay off, you've got $100,000 worth of equity. Most banks will lend you 80 percent of that, provided they ascertain that you will be able to handle the monthly payments. Generally, you pay no points or finance charges, interest rates are relatively low, and the interest on the first $100,000 is tax deductible.

With their minimal paperwork, flexible payments, and favorable interest rates that may be tax-deductible, home-equity loans are an excellent source of remodeling money. But proceed cautiously because you're putting your greatest asset on the line. The American Bankers

Association suggests that you ask lenders the following questions before you sign on the dotted line:

- What is the annual percentage rate. If an introductory rate applies, how long will it last? How is the interest rate established? How often is it adjusted?
- How high can the interest rate climb each year and for the life of the loan? Is there a limit on low it can go?
- What up-front costs are involved? Are there annual maintenance fees or other fees? For a home-equity line-of-credit loan, is there a minimum draw requirement?
- What is the length of the loan and what are payment terms and options? Will there be a lump-sum payment at the end?

With a home-equity loan, you'll receive a check for the entire amount you borrowed and a book with monthly payment stubs; interest will begin accruing right away. With a **home-equity line of credit,** which works something like a checking account, you can withdraw the money as you need it and interest will be charged only on the amount you draw out. You need to be disciplined here and resist the impulse to use this stash for a vacation or a new car.

You might also get your remodeling money by **refinancing your mortgage.** With the loan you acquire, you pay off your existing mortgage and use what's left over to finance your remodeling. You're in a good position to do this if your house has substantially increased in value since you bought it and if the current interest rate is two or more points lower than the rate that accompanies your existing mortgage. Otherwise, the closing costs attendant upon taking out a new mortgage will make this sort of loan too expensive; in that case, a home-equity loan would make more sense.

If you are planning a major overhaul, you may be able to get financing through a **home-improvement construction loan.** Once you have hired a contractor, the two of you work together to complete an application and construction-completion schedule for your lender of choice. When the loan is approved, draws against it are released according to the prearranged schedule. After each stage of the renovation is finished to your satisfaction, you sign a completion certificate, which the contractor submits to the lender. The lender pays the contractor and you start repaying the loan. Important: Ask your lender to verify that the contractor has paid all subcontractors and suppliers before it releases funds for subsequent phases of the project.

Master This Mini-glossary

Construction loan. Financing that covers building or renovating a house. Once construction is complete, you must obtain a permanent mortgage to pay off the loan.

Fixed-rate loan. In this type of financing, the interest rate remains constant throughout the life of the loan.

Adjustable-rate loan. With this type of loan, the interest rate varies periodically according to a predetermined financial index.

Points. The cost of obtaining a loan, with each point equaling 1 percent of the total amount borrowed. Lenders often compensate for a lower interest rate by charging the borrower points.

Fees. The one-time cost of a loan, such as appraisals, lawyer's expenses, points.

Annual percentage rate (APR). The total annual charges for a loan. The APR includes the interest rate charged on the loan plus all other fees, incorporating all of these costs into one figure as a percentage of the total loan. The government requires lenders to make the APR available to borrowers.

Even if your equity is low, you might have some luck with loans offered by the Federal Housing Administration (FHA), a division of the Department of Housing and Urban Development (HUD) or the Federal National Mortgage Association (also known as Fannie Mae). The Title I program offered by the FHA, for example, sponsors some home-improvement loans with 15-year payback periods that reduce monthly payments substantially. The FHA's 203 (k) program is another attractive prospect. Check with your accountant, financial adviser, or bank to see if you qualify for one of these loans (or see Chapter 7 for how to reach HUD or Fannie Mae directly).

Because the government guarantees full payment to the local lender even if the homeowner defaults, these loans are sometimes easier to obtain than conventional ones. But be prepared to be patient. Securing a HUD loan takes a long time, requires much paperwork, and involves many restrictions and regulations. In addition, government interest rates may be higher than those charged by lenders in your community.

Other sources of remodeling money may work for you: taking a loan against the value of your stock portfolio, borrowing from a life-insurance policy with a cash value, tapping into your 401 (k) retirement plan or profit-sharing fund.

WHERE TO START LOOKING FOR A CONTRACTOR

Now that you have done your homework, established a budget, figured out financing, and decided to proceed with your project, it's time to begin looking around for a contractor. The next chapter takes you step-by-step through the nuts and bolts of interviewing serious contenders, checking references, and making sense of bids. Here we point you in the right direction with some general advice about where to start looking. The best approach is to get a roster of names from people who have little or no vested interest in your hiring a certain individual.

News media

Magazine and newspaper articles, news briefs on radio or TV programs, and trade associations all offer advice on where to find a contractor. Their suggestions are generally fairly sound but often do not go far enough. Let's see how other traditional sources measure up.

The Yellow Pages

Use this often-suggested source only as a tool to learn what kind of work a potential candidate does and where he or she is located. The telephone directory does tell you who is in business in your area, and display ads probably tell you what a contractor's specialties are. If an ad says "commercial only," for example, you have saved yourself a phone call. If you're adding a master suite to your home, you don't want a firm that builds shopping centers.

Legitimate and shady contractors advertise side by side in the Yellow Pages, so you must not confuse a listing with a recommendation. Anyone can take an ad and sprinkle it with words such as "dependable," "experienced," "skilled." The listing agencies do not verify claims of this kind. Beware; your worst nightmare could be awaiting you in these listings.

The real-estate and home-improvement sections of your local newspaper probably also offer contractor ads.

Lumberyards, home centers, and other suppliers

This source of contractors' names should also be viewed with a certain amount of suspicion. The names that you will be given when you ask for good contractors in the area are customers of the store, so this is not an unbiased source. You should take these recommendations with a grain of salt, at least until you have checked further.

These sources do have *some* value, however. Store employees are not likely to recommend someone they know does shoddy work or is constantly in arrears, because they don't want you—also their customer, after all—coming back to them in a rage. You can probably tell with a few well-placed questions whether a certain contractor is busy and what the store employees think of him or her.

One store manager of a large lumberyard—home center in upstate New York says that he and his employees are constantly asked for contractor recommendations. "We give people a half-dozen or so names," he says, "but we won't recommend any one person in particular. We say, 'Here's a list; don't assume the first name on it is better than the others.' Anyone can buy a level, a hammer, and a pickup truck, and call himself a contractor. We urge people to get references from previous customers."

The how-to-find-a-contractor advice often suggests asking home centers if a particular contractor pays bills on time. In real life, however, most store employees are reluctant to answer that question. "People ask," says the store manager, "but we don't answer. We feel that it's confidential information."

Business organizations

Other possible sources for contractors' names are the Better Business Bureau, chamber of commerce, and construction-loan departments of banks, and are additional possible sources of contractors' names.

Subcontractors

Electricians, plumbers, roofing and siding contractors, cabinetmakers, and floor installers all probably work on a regular basis with several of the contractors in your area, and they can be sources of valuable information.

When you ask them for recommendations, begin the conversation by saying that you are planning a remodeling and are looking for a good general contractor. Then ask: Is there any g.c. you think does particularly good work? If a name is forthcoming, ask: Have you ever worked with this person? On what basis are you recommending him? Did he pay you on time?

Your architect or designer

If you are using an architect on your project, you will very likely receive contractor recommendations from him or her. Interior designers often have favorite contractors too. Hiring one of the contractors that the designer or architect recommends might seem as though it would save time and trouble. However, proceed as you would with any other name that you have been given: interview the candidate, check references carefully, and talk to previous clients. Do not abdicate responsibility for this important decision to your architect or designer.

If, on the other hand, you have an ongoing relationship with the design professional or have come to trust him or her through the hiring and design process, you can probably go ahead and take the recommendation. You do not have to use the people your architect recommends, of course, but it would wise to at least consider them. Says architect Steven House, "We have a track record of working with builders and contractors; we're almost like a clearing house. We know who has skill, integrity. We probably know more about these contractors than the homeowners could learn in the hiring process."

The Internet

See Chapter 7 for a list of a few web sites that will give you the names of contractors in your area. Don't assume from these listings that the contractors are skilled and dependable, although some sites, such as www.Improvnet.com, claim to have checked these people out for you. Even so, it's smart to do the usual research yourself.

Trade associations

Several organizations, most notably the National Association of the Remodeling Industry (NARI) and the Remodelors Council of the National Association of Home Builders (NAHB), will list member contractors in your area. The fact that these people have paid dues and fees for membership does not imply skill and dependability of course. As N'ann Harp of Smart Consumer Services puts it: "Trade associations don't have a reputation for policing their members 100 percent or stringently sanctioning them for poor professional behavior." But the fact that a contractor has taken the trouble to join a professional trade association does demonstrate a certain seriousness and sense of responsibility.

In addition, certification letters after a contractor's name indicate that he or she has passed a professional course of study provided by one of these associations. The NARI program offers three levels of certification: CR (Certified Remodeler), CRS (Certified Remodeler Specialist), and CLC (Certified Lead Carpenter). The Remodelors Council of the NAHB provides a CGR certification (Certified Graduate Remodelor).

Local architects

Even if you are not using the services of an architect on your project, some of them may still be willing to give you the names of contractors they work with often. If the architect has a thriving, longstanding business, that in itself is a good recommendation. He or she couldn't afford to work with contractors who did a bad job. "I would definitely give out the names of contractors I work with," says architect Steven House. "Although quite honestly, I probably wouldn't give the name of someone who's currently working on one of our projects because that would conflict with my client's best interests."

Real estate firms

Whether or not you are making immediate plans to sell your house, local real estate brokers or sales agents may be willing to give you contractor recommendations. Because they frequently see the results of home-improvement projects and often help sellers arrange to fix up their houses before putting them on the market, they are probably quite knowledgeable about who is doing good work in your community.

Friends and neighbors who have had work done recently

This is a good and usually untainted source because you can actually look at the job that was done and examine its workmanship. Also, people in this situation are usually willing to go into some detail about why they liked—or didn't like—the contractor. Hearing that "so-and-so was wonderful; we were very pleased with his work" is very encouraging, but you'll also want some specifics, such as Did the crew always show up on time? Did the job stay on schedule? and so forth.

One reason friends, relatives, and neighbors are such a good source of recommendations, says New York City contractor Mike Vella, is that the contractor has a reputation to uphold. If he did a great job for Aunt Helen and she has praised him to the skies, he's going to want to perform equally well for her friends and neighbors.

But going to friends and neighbors is also a little tricky. A recommendation from them doesn't tell you much unless you ask the right questions and unless the project is comparable to your own. A contractor who built a deck for the Smiths or installed a new kitchen sink for the Browns will not necessarily do a good job on your more ambitious project.

References from friends and relatives can be especially fraught with peril if the names they give you are friends or relatives of theirs. Check these recommendations very carefully or find a gracious way to avoid them altogether. Hiring your best friend's cousin or your next-door neighbor's uncle can bring big headaches if things don't work out well. If you run into serious problems and have to fire the person or institute legal proceedings, it will very likely ruin your relationship with your friend or neighbor. For that same reason, be careful about hiring your own relatives.

Projects in progress

It's perfectly acceptable to stop by ongoing jobs that are similar in scope to your own and observe the work. Does it seem to be getting done in a orderly manner? Is the job site clean? Does the quality of the work seem to be good? You can also ask the homeowners—preferably when the contractor is not there—whether they would recommend him or her for the project you are doing.

"The bottom line," says N'ann Harp, "is that it almost doesn't matter where you find a contractor. But it does matter how carefully you check each candidate out."

DO LICENSES MEAN ANYTHING?

Licenses don't mean much, say most experts. But paradoxically, it is important that you make sure your contractor candidates have licenses, and that you ask to see the licenses, record the registration numbers, and check to see that they are current.

According to N'ann Harp, "Many contractors are of the opinion that a license is a worthless piece of paper that merely gives the local government another reason to collect money from businesses. I tend to agree." After all, says Harp, in most states the licensing procedure typically doesn't involve a skills test. (An exception to this is California, which requires four years as a journeyman, foreman, supervisor, or contractor in the trade for which he or she is applying before a license can be issued.) "Getting a driver's license at least requires that you demonstrate basic skills behind the wheel in the presence of an examiner," she adds. "Getting a contractor's license does not require any demonstration of skills. Homeowners would be wildly mistaken to imagine there is any workmanship reassurance to be taken from a contractor's license." Possession of a license does, however, indicate that the name of the owner and operator of the business is on record somewhere, which might be helpful in the event of an insurance claim or lawsuit.

Tom Philbin, the author of dozens of books and hundreds of articles on the home-improvement field, doesn't put a whole lot of stock in licenses either. They may mean only "that a contractor paid a fee for being registered or licensed and proved that he had liability insurance.

In few states do they measure competence or honesty," says Philbin.

Although you should not overestimate the fact that a contractor possesses a license, you could allow yourself to be somewhat comforted by the fact that he or she took the time and trouble to get one. It may indicate a sense of responsibility and a professional attitude toward the job and an interest in running a legitimate business. At the very least a license tells you that a contractor meets the minimum standard for his state; and in some states, it may also tell you that he or she has passed a competency exam.

Oddly enough, although a contractor's license doesn't tell you much about the abilities and character of the person, the lack of one could be significant and you should view it as a red flag, say most experts. Hiring someone who is not licensed can become a real problem should a dispute arise or should injuries or damage to property occur. Your local consumer affairs agency will have no clout with an unlicensed contractor because it can't suspend or cancel his license or fine him. And you will have no access to the plan that some states offer to refund money to clients when contractors can't do so (licensed contractors contibute to the fund to cover such contingencies).

So before you compile a list of possible contractors, find out if they are licensed. And get in touch with your consumer affairs bureau to see if your state has a refund plan. In some states you must file suit in civil court before you can be awarded a judgment; in other states you simply apply to the local consumer affairs agency. Says Tom Philbin, "The great thing is that it means the money is out there even if the contractor has fled the scene or has gone out of business and is insolvent." But, he warns, in no state can you collect if the contractor isn't licensed or registered. Another drawback: an unlicensed contractor is probably an uninsured contractor. And that could cause you a great deal of trouble. (There is additional information on insurance issues in Chapter 3.)

It is possible of course that an unlicensed contractor is perfectly honest and competent and will do a great job for you. But as a careful consumer, you must question why he or she didn't take the trouble to apply for the piece of paper, pay the fee, and if required, take the test. Don't let the lack of a license become a deal breaker; but if you hire a contractor without one, do protect yourself in other ways—keep your own liability insurance up to date, check references carefully, and prepare a solid contract.

To learn whether your state requires licenses or offers refund plans, refer to the state-by-state list in Chapter 7. Then call the appropriate agency to discover whether your candidates' licenses are bona fide and up to date.

SCAMS AND CON MEN

Most contractors are honest and hardworking. However, you do need to be on the alert for that occasional bad apple. N'ann Harp likens home improvement to a high-stakes game. On one team are the Homeowners who put their nest eggs at risk, she says. "On the other side, listed neatly and alphabetically in the Yellow Pages, are the Contractors, good, bad, and mediocre, all sporting nearly identical ad pitches." But, Harp continues, lurking on the sidelines are the worst of all, the Fly-by-Nighters. They sweep through a neighborhood by telephone or in person, promising great deals. Trouble is, they tend to disappear with your money without doing much, if anything, for it.

Unscrupulous contractors and fly-by-night con artists often target senior citizens, believing that these elderly homeowners have long since paid off most of their mortgages and are likely to have the funds available to finance home improvement projects. But it's not just senior citizens who get taken, says Harp. "Anyone who's too busy to do their homework is putting hard-earned cash at risk." Protect yourself, she suggests, with a couple of lessons from Smart Consumer Services' "Homeowner Survival Kit."

Buyer Beware Rule No. 1

Never accept an offer for home repairs from someone who rings your doorbell or telephones and starts off with, "We've been doing some work in your neighborhood...."

Say you have a slightly crumbling blacktop or gravel driveway, and one fine day a paving-company truck pulls up in front of your house. "Hi!" says the pleasant young man who knocks on your door. "We just finished paving a driveway in the neighborhood [he waves vaguely up the street] and happen to have a bunch of leftover material that we hate to have to take to the dump. I couldn't help noticing that your driveway looks kind of rough. We could pave it right now for a real low price. Otherwise, we'll have to throw out this leftover material."

He quotes a price. It *sounds* good. But how would you really know? Have you priced driveways recently? But you accept the offer and they pave the driveway. It looks good for a couple of weeks, until grass starts sprouting through the pitifully thin layer of tar that was spread.

When you make a few calls to see what local companies charge for driveways, you discover you paid a very high price for a very shoddy job.

ImproveNet.com, an on-line remodeling information service, describes another typical scam. One day, a retired couple heard banging and hammering on the roof of their Minneapolis home. The went outside, looked up, and asked the man on the roof what he was doing. He said he worked for the company who had originally installed their roof and he was doing standard yearly maintenance work. He hammered down some more shingles (or pretended to), then presented the owners with a bill for a thousand dollars. He was down the road before the ink on the check had dried.

Buyer Beware Rule No. 2

Know your rights. Harp cites another typical high-pressure operation, "today's special." A replacement window salesman—very polite and nice, just like the driveway scam artist—sits in your living room and shows you pictures of his line. Then he uses your phone to call his office for his boss's approval to give you a one-time, tonight-only special low price. Impressed, you sign the contract.

The next day, you have second thoughts but figure it's too late; after all, you've signed a contract.

Actually it's not too late. According to federal law, you have a three-day cooling-off period, also called the Right of Recision period, to withdraw from any non-emergency remodeling or repair contract over $25. You can call the company, tell them you want to think it over, and formally withdraw from the contract. You might also write a brief note to the company, repeating in writing what you said on the phone.

In short, whatever the size or scope of your project do not be pressured into "we were in the neighborhood" or one-time only offers. ImproveNet also suggest that you avoid anyone who wants you to make an immediate decision, demands payment up-front, or hesitates to give you clear references.

Heads Up: Watch Out for These Bad Apples

According to ImproveNet, home-improvement scams can be divided into three categories: Rip-Off, Low-Ball, and Shoestring.

Rip-Offs are out-and-out crooks, like the con artists described above. They will do anything, promise anything, to get your money. Then they'll disappear without doing any work.

Low-Ball con men may be dishonest or poor business managers or both. They quote very low bids in order to get the job; then come up with costly change orders so that they can make the money they needed in the first place. It may be deliberate, but often it's the result of incorrectly bidding a job. Either way, the homeowner gets taken.

The *Shoestring* contractor is struggling, having trouble staying ahead of the money. He or she does not intend to rip anyone off, but the hapless clients of the Shoestring may end up spending more than they intended to, and the Shoestring contractor may lose money too. Shoestrings—good at building, bad at business—make up a significant segment of the remodeling field.

A CAUTIONARY TALE: CONTENDING WITH THE SHOESTRING CONTRACTOR

Sure, there are some unscrupulous contractors out there, con artists who want to take your money and run. However, you're more likely to encounter contractors who do shoddy work, and others who do excellent work but are bad at business and worse at effectively scheduling their jobs. Either type, although basically honest, can cause just as much havoc as an out-and-out con artist can. Sad to say, many contractors of this type go from job to job doing poor work or causing financial chaos because potential customers don't check them out carefully or talk to previous clients.

There is such a contractor lurking in the wilds of western Massachusetts, possibly at this moment making some unsuspecting homeowner's life miserable, possibly using the same tactics he employed with Fiona and Bernardo (not their real names).

Fiona and Bernardo made some very human and very typical mistakes when they hired this particular contractor. We'll call him Joe. They didn't check his references, they gave him the benefit of the doubt, and they gave him a little too much money up front. They didn't get hurt as badly as they could have, but theirs is a sad story nonetheless.

Owners and operators of a real estate firm in a small town, Fiona and Bernardo were a bit surprised when Joe came into their office and asked to be put on the list of contractors the firm occasionally recommends to home buyers. They were surprised because, says Fiona, "In a small community like ours very little is secret and we knew that there had been complaints about his performance in the past." Even so, she and Bernardo were impressed that Joe admitted that he had had some troubles, and said that he now had his act together. "We knew him and thought he was a nice guy," says Fiona. "He sold himself well, and we wanted to give him a chance."

They gave Joe a chance by hiring him to do some work on their house—expand the deck, build a staircase, and replace creaky railings and uprights. They didn't ask for references because, explains Fiona, "We knew they'd probably be negative, given the troubles he'd been having."

The job went smoothly, Fiona recalls. "He was very professional, showed up on time, and did beautiful work." Based on this positive experience, the couple hired Joe for another, bigger job—transforming an old store into a satellite real estate office that was due to open in a few months. They did not ask for references about Joe's ability to handle the bigger job. After receiving a professional-looking, written estimate from him, they did not seek other bids. "His price was about what we thought we'd have to pay," says Fiona.

This job started out fine, but soon signs of trouble appeared. Joe didn't show up on time every day. Then, now and again, he didn't show up at all. When Fiona and Bernardo spoke to him, he was always apologetic and assured them that everything was now under control. "Not a problem," he said, over and over. "It was his mantra," says Fiona. Things got worse. Empty beer cans showed up on the job, Joe missed more and more time, then did fast, sloppy work to catch up. One day Bernardo heard some disturbing things about Joe: he had a drinking problem, he was taking on many other jobs, and was in trouble with all his other clients for taking their money and not doing the work. Concerned, the couple had a talk with him, but it didn't help. He continued his absenteeism and shoddy work, and finally they fired him.

Out a couple of thousand dollars and relying on two hastily hired contractors to finish the work on time, Bernardo and Fiona reflected on where they went wrong. One, because they liked Joe, they were too willing to give him a chance. "Maybe we were caretaking," says Fiona. Two, they didn't talk to previous clients or ask to see previous jobs. If he had truly cleaned up his act, there might have been a few satisfied customers to attest to it. He did good work for them on their deck, but they didn't know if he could handle a bigger job. Three, they didn't have any fall-back plan for the possibility that he would not show up. A carefully constructed contract could have included stipulations and penalties for that. Four, they weren't tough enough. They did confront him every time he failed to show up, but they also accepted his excuses every time. Five, they didn't ask anyone else to bid on the job. "The whole thing seemed to be working to our advantage," says Fiona. "We needed someone right away for a deadline project; he had just done a nice job for us on our deck; and he was available. We didn't want to take the time to look further.

"Success can be hazardous to many small, independent contractors," says N'ann Harp. "That is the most significant detail in this cautionary tale. Just because a small business can handle a couple of jobs at once does not mean that they can, or should try to, handle three or four. But they just can't say no.

"So one of the most important questions a homeowner can ask a potential contractor would be, 'How is your business going?'" Harp suggests. "And if you get an answer like, 'Great! I have more work than I know what to do with!' you should probably translate it as 'I'm trying to juggle too much and I have no delegation skills.' Or 'I underbid the last job and I'm trying to cover those costs with deposits from new jobs. But to be sure I get the new jobs, I'm low-balling them too.'

"Ironically, the more successful the small operator becomes, the greater the risk for problems," Harp concludes.

A SUCCESS STORY

Not long ago Jason and Leslie James (not their real names) remodeled the kitchen in their 35-year-old northern Virginia house. Why was it so successful? Because they did several important things right.

First, they chose a contractor whose work they had seen and who was highly recommended by people they knew. “Good friends of ours live right down the street in a house that’s very similar to ours,” says Jason. “When they had their kitchen redone, we watched closely. We saw it under construction, and we spent many hours in the finished room six months and a year after the job was completed. We knew that the remodeling proceeded in an orderly way and that the work held up well. We were impressed that the contractor was willing to come back months later to make a few minor adjustments.

“All of these things influenced our decision,” Jason says. ”And we factored something else in: our friend’s nature. She is very picky, one of those people who diligently researches every choice and decision she makes. If she thought this contractor was worth trusting her kitchen to, we figured she was probably right.”

Another smart thing the Jameses did was to get bids from other contractors, even though they were pretty well sold on the first one. “We did it mostly to see if our guy’s bid was in line with others,” says Jason, “and also to get a sense of how other professionals saw the project. We were pleased that our guy came up with ideas that we hadn’t thought of; the other contractors hadn’t thought of them either. His price was fair, he was personable, we sensed that he wanted to do it right, he showed up on time for meetings. That was enough for us.”

Well, almost enough. The Jameses also drew up a detailed contract that put everything in writing. Says Leslie. “It was a way of making sure we agreed on what was going to be done. If we hadn’t gotten it all in writing, we might have been fuzzy about something six weeks down the road. This way we were all on the same page all the time.”

three

Interviewing the Professionals

At this juncture, having followed all the good advice in Chapters 1 and 2, you have probably developed a preliminary list of contractor candidates and are ready to begin the decision-making process.

Don't hurry. Finding the right contractor requires careful attention and time. You will be investigating the skill, experience, work ethic, and financial stability of each candidate. To hurry the investigation—or to make a decision for the wrong reasons—is to invite big and expensive problems.

Before we proceed with guidelines for the decision-making process, we should address the indisputable truth that the responsibility for home-improvement woes often falls on the consumer. We are often disappointed by contractors because we have unrealistic expectations or we choose the lowest bidder or we fail to educate ourselves about the scope and likely cost of our project. Unfortunately, we also frequently fail to carefully check out our contractor candidates and don't bother to follow even the simplest guidelines for making a choice. And what happens to projects for which all care has been abandoned?

Sometimes, things do turn out all right (more about that later), but almost always, they don't.

"Checking references is the smartest thing you can do to protect yourself," says N'ann Harp of Smart Consumer Services. It's also the most time-consuming, which may be why people so often zip through the process too fast, or worse, sidestep it altogether. But checking references could prevent a catastrophe, says Harp. "The chances of recovering a dime of money lost to a fraudulent, abusive, or incompetent contractor through a lawsuit or criminal action are slim to none. Hundreds of thousands of dollars of hard-earned money could be saved if every homeowner took this important step."

And yet we balk at doing it. Why is this? We all know that a major remodeling job will cost money and that the people executing it will be our part-time housemates for a while, maybe months. We all have heard horror stories from friends and acquaintances and occasionally read about rip-offs of various kinds in the daily press. Why are we not more careful?

"Because people don't know what to ask," says Keith Slater, an insurance agent whose Tualatin, Oregon, firm handles business for about 500 contractors of all sizes. "If a guy sounds nice, they'll hire him without any verification. It seems to be human nature." Carol Little, a psychotherapist with a private practice in Delhi, New York, has some other theories about why we fail to protect ourselves against potential contractor problems. "People in general, whatever their background or educational level, often have some trouble believing that they are deserving. This lack of confidence seeps into all areas, including hiring others to do work for us, which we may feel very uncomfortable about. Many of us fail to stand up for ourselves because we don't feel we have the right to question the skills and abilities of another."

Another problem is the naive way that many of us look at so-called experts, says Little. "Of course we have an absolute right to question anyone to whom we are going to be giving money, but we get it backwards. We want the 'experts' to approve of us rather than requiring them to earn *our* approval."

And once we enter into a relationship with a contractor or anyone else we have hired, we let ourselves be pushed around. "We need to be assertive," says Little, "to be able to ask for clarification or to slow the process down if we feel we're being swept along too fast. If we feel intimidated enough, we may even zone out and stop listening, and then we can end up in big trouble.

"I've noticed," says Little, "that people who have money tend to be more careful about protecting their resources and about hiring the best people for the job. They're accustomed to hiring people to do all kinds of work for them and have no problem asking tough questions and insisting on safeguards."

So what's the answer? Pretend you're rich, convince yourself you have a right to get the best person for the job, and get a little assertiveness coaching. There are many books on the subject, but Little thinks *When I Say No, I Feel Guilty* by Manuel F. Smith, which deals with both business and personal encounters, is one of the best.

Cynthia Smith, a psychiatric social worker who is employed by a New York City medical center and has a private practice as well, thinks we fail to be careful because we want to be taken care of. "Something that must be addressed is the undeniable fact that many people ignore all valuable advice about making a careful decision and hire the first and only contractor they talk to because they feel so comfortable with him or her or that they really believe the careful way is not necessary.

"Somewhere within many of us there is a hope that the contractor we like and trust will care about us and do a good job for us," Smith theorizes. "The trouble is, unless we come across a truly skilled and ethical professional who also has a need to take care of us, we're in trouble.

"Sometimes the reasons that we fail to be careful are rational," Smith continues, "such as being especially short on time or believing that we will lose a promising contractor if we don't hire him or her right away." Other reasons, she thinks, are less defensible. "Some of us may feel dependent on the contractor—and on other 'experts' such as lawyers, doctors, accountants—and along with it, hope that they will take care of us."

However temporary and fleeting they are, our dealings with a contractor constitute a relationship, says Smith, and all relationships present the same basic issues. If assertiveness is a problem for us, it will rear its head with our home-improvement professional as surely as it does everywhere else. We may find it difficult to state our needs clearly; we may be anxious about upsetting or alienating him or her. "To guarantee that the prospective contractor will like us and do a good job for us, we think, we must not take a chance on making him or her mad by asking tough questions about licenses or insurance or previous performance. And sometime we go so far as to not insist on a contract because to do so will hurt someone's feelings and give the impression of mistrust. In these interactions we may feel as though we don't have any power, which is another basic relationship issue. We don't see the professional as being glad to have our business: we see ourselves as lucky to have him or her. Of course the reality is that we do have power, but we fail to use it." Also, says Smith, when we get a good feeling about one prospective contractor, a feeling of trust and rapport, say, we want to go with it. We don't want to hear those other, more careful, voices that tell us we should check him out or look further. We don't want to find any problems because then we may have to deal with them assertively.

What about this feeling of trust and rapport? Is it important? N'ann Harp doesn't think so. "Sure, choose someone you like and get along with," she says. "but as long as you have checked out the contractor, developed a good contract, and ascertained that you can work together, it really doesn't matter if you like the guy." OK, so you don't have to fall in love, but neither should you choose someone you're sure to have problems with. Listen to your own inner voice, and pay attention to your personal triggers. Right or wrong, healthy or neurotic, you probably should not hire someone who intimidates you or reminds you of a demanding boss or a fickle ex-sweetheart. You might want to work those issues out in therapy, but a construction site in the midst of your own home is not the place to engage on a daily basis with someone you find particularly difficult. A consumer booklet issued by the National Association of the Remodeling Industry (NARI) expresses it this way: "Ask yourself if you would mind greeting this person at the crack of dawn or running into him or her after a long day at the office. Does he or she seem like the type of person who would be easy to work with and who would be careful with your belongings? Face it: This individual will be a part of your home life for the duration of the project."

START MAKING CALLS

At this point you have a good idea what the scope of your project will be, a rough idea of what kind of budget you can handle, and a list of likely candidates for contractors, gleaned from some of the sources we recommended in Chapter 2. From those sources you should compile a list of 8 to 10 names. Sounds like a lot? Once you realize that the next few steps you are going to take may eliminate some of those names, leaving you with only a handful, 8 to 10 names or even more will not seem like too many.

The rest of the process consists of research and weeding out of questionable candidates, then carefully interviewing—and doing more research and weeding—of the ones that remain.

First, find out if any complaints have been registered against your candidates, and if possible, what kind of complaints. You can do this by contacting two helpful organizations, your state's attorney general's office and your local Better Business Bureau (BBB). Your attorney general's office may be able to tell you if any disciplinary or legal action has been taken by a governing or licensing body against any of your candidates. Not all attorney-general offices will release such information, but you can give it a try. You could also commission your own attorney to search out this information for you, at a cost of at least a couple of hundred dollars.

Your state attorney general's office also includes either a person or an entire department that handles consumer affairs, investigates complaints, and alerts citizens about frauds. With seven regional offices throughout the country, the Federal Trade Commission (FTC) also handles consumer affairs. The FTC can keep you up-to-date about contractor scams in your area and can refer you to the appropriate agency if you need help with the hiring process. (You will find state-by-state listings of consumer affairs offices and Better Business Bureaus in Chapter 7.)

None of these worthy organizations can tell you whether or not a certain contractor is honest, dependable, or skilled. They can, however, give you valuable information that will help you draw your own conclusions. Or, as Ron Berry, senior vice president for bureau affairs for the Council of Better Business Bureaus, puts it: "We can't predict the future but we can give you the benefit of the experience other people have had in the past.

"There's no question that people should be a lot more careful about hiring general contractors, or contractors of any kind," says Berry. "A look at our complaints reveals a clear pattern: had consumers done their homework and been better informed, they never would have chosen the particular company or contractor that they ended up complaining about. We also find that consumers who registered complaints against contractors didn't really understand the remodeling industry or the project they were embarking on."

In other words, if you avail yourself of the inquiry services of the BBB in the first place, you probably won't need to call the complaint department in the second place.

The BBB divides the records of contractors and contracting companies into three categories: satisfactory, unsatisfactory, and neutral. A satisfactory record is earned by businesses that have been in existence for at least a year and have accrued no consumer complaints, or only a few, all of which have been addressed and resolved. An unsatisfactory BBB record is given to a firm or individual that has received many consumer complaints and has not resolved them, or has resolved them only to have the same sort of problem occur again and again. This, says Berry, "reveals a pattern. Some companies will address and resolve individual complaints because they know they have to in order to stay in business. But because the same kinds of complaints keep coming in, we can see that they have not changed their ways." The neutral category describes a kind of gray area wherein a company or contractor may be too new to have established a record of any kind or may have a mixed record, with some complaints resolved to the customer's satisfaction and some not.

BBB complaints are most often resolved between contractor and homeowner, with the contractor addressing the problem and taking care of it to the customer's satisfaction. But occasionally resolution requires an arbitration hearing, which is administered by the BBB and community members who have been trained in the arbitration process.

If any of your candidates has a satisfactory record with the BBB, keep this name on your list and proceed. A neutral rating, says Berry, is not as comforting as a satisfactory one, but neither should it scare you off immediately. If the contractor in question has been in business a long time, handling many jobs each year, he or she is bound to have tangled with someone. A complaint does not mean that the contractor is wrong. It's merely an indication that you should follow up by asking your candidate for his or her side of things. If the rating is unsatisfactory, scratch that name off your list. No point in looking for trouble. As a further safeguard, ask the BBB or consumer-affairs representative whether the number of complaints the candidate has accrued is more, less, or about average for the general-contractor trade.

Your next step will be to make preliminary calls to all the people who remain on your list of candidates. Start calling a minimum of three months before you want to begin your remodeling project. If you have noticed a lot of remodeling activity going on in your locality, start making your calls six months or more ahead of time. In a really busy market some contractors can be booked as far as a year in advance. For a small job, such as adding a deck or installing a tile floor in your bathroom, two months may be adequate lead time. By the way, you will avoid delays and many rounds of telephone tag at this stage of the process if you try reaching your candidates either before eight in the morning or after five at night. Contractors generally don't spend much time in their offices.

Tell each person on the list where you got his or her name, briefly describe your project and the dates you want to have it done, and ask if he or she is interested in taking a look at it. This process will probably eliminate a few more names from the list. Some contractors may consider the job too big, or too small; some may not be able to meet your time requirements; and some, sad to say, may never call you back. If you get a yes from any of these people, proceed with a few more important questions: Are you licensed or registered in this area? Do you carry insurance for liability and workers' compensation? Immediately eliminate those who don't have licenses or insurance, especially if your project will be a major one. You can do this smoothly with a comment like, "Thanks. I appreciate your help. I'll have to do some more thinking before I decide to go ahead," or something equally noncommittal. (Some states do not require licenses; for your state's requirements, check with the appropriate regulatory agency listed in Chapter 7.)

Go one step further with those who respond that they *do* have licenses and insurance—ask them to bring photocopies of those documents (or the registration and policy numbers of said documents) when they come to talk to you about the project.

A word should be said here about hiring a remodeling contractor in a small town or a rural area. In such locations, it may be customary and even preferable to ignore some of the sage advice you have been reading. In some areas, few of the local contractors do things by the book and may even be offended if a homeowner asks for references and an ironclad, lawyer-approved contract. Such was the experience of Graham Schelling, a freelance writer and editor, who bought a country house in rural Wisconsin several years ago. The house needed a lot of work and Schelling asked a neighbor, someone he had known when he lived and worked in Minneapolis, to suggest a contractor. He didn't ask for other references, didn't check previous work, didn't verify licenses or insurance, and never drew up a contract.

Vern, the man he hired in a such a cavalier fashion, started with the roof, did a good job, and has been working on the old farmhouse ever since, renovating it bit by bit. He tore down old plaster walls and rebuilt them with drywall, added a new bath, remodeled an old one, transformed an outbuilding into an office for Schelling, and insulated and rewired the whole house.

Because the work was done piecemeal there was time for a relationship to develop. "Now," says Schelling, "he will see something that needs correcting and tell me about it. 'I really don't like the look of those old basement windows. They're about to fall out,' he said one day. 'OK, let's replace them,' I said. Now the basement is brighter and airtight. We've come to trust each other so much that he will occasionally fix a small thing, like a faulty porch light, without even asking." There are two drawbacks, though, says Schelling, "He's slow because he works by himself, and when there's heavy lifting to do, I have to help."

Why did Schelling approach this relationship with so few safeguards, feeling that one reference was enough? The reference carried a lot of weight, he says. "But the real reason was I liked him and trusted him right away." Schelling also sensed that to ask for references or a contract wouldn't go over well with Vern. "I had a strong feeling that business on a handshake was the way things are done here and that my best bet was to go with the flow."

So if you're new to the small town or rural area where you live, try to find out what the local customs are before you start asking for big-city safeguards that may offend someone. It's possible that you will ignore conventional wisdom, hire the first contractor you meet (after all, there probably aren't a great number to choose from), and come out of it just fine. But, unlike Schelling, don't even dream of giving up the protection of a contract. It may be a simpler document than the one you'd sign with a high-powered contracting firm in a metropolitan area, but you must have something that guarantees the basics, the ones we will discuss in Chapter 4.

Here's one way to go about it, as suggested by N'ann Harp. "Smart homeowners would take the negative-option approach," she says. "Something like this: 'You wouldn't mind if I put together a simple statement, describing what you'll be doing and the price, would you? Just so we're sure we both understand each other?' And, using a sample contract, proceed. A comfy, cozy lifestyle in the country notwithstanding, you can get into all kinds of trouble without legal protection."

THE INTERVIEWS

Ideally, interviews should take place at the job site, which in most cases will be your home. Schedule appointments with the candidates by asking them to come to your house at a certain time. Experienced contractors will assume that you are meeting with several people, but there is no point in setting up an adversarial atmosphere by scheduling the meetings so close together that the competitors bump into each other. If possible, schedule the meetings on separate days, or at least an hour apart.

After a few interviews, or even after the first one, you may be so impressed with someone that you want to hire him or her on the spot. Should you keep interviewing even though you have pretty much made up your mind? Yes. The next people you talk to may have ideas about a better way to approach the project, or how to cut costs significantly or make the result more pleasing aesthetically. Hearing these thoughts may cause you to change your mind; but even if you end up hiring the person you liked so much in the beginning, you will have gained a helpful new perspective on the project. Also, it's too soon to make up your mind, no matter how impressed you are. There's a little more checking to do, the results of which may make you glad you didn't hire someone hastily.

To create a friendly, relaxed atmosphere for everybody, sit at the kitchen table or in comfortable chairs in the family room. Make the interview friendly but businesslike. It is a job interview, after all, not a social call. Make sure you take control of the meeting from the very first and don't let any of the candidates take over with a sales pitch. If you're in control, that won't happen. But do keep in mind that this is not an interrogation. If you have any lingering negative thoughts about the ethics or trustworthiness of contractors in general, now is the time to get rid of them. Otherwise, they will surely affect the interview. Yes, there are some disreputable people out there; but your careful attention to the hiring process will eliminate them—indeed it may already have eliminated a few bad apples. So there's no need to poison the air and the interview with negative thoughts about the person who's sitting in your living room hoping to get the job. The purpose of the interview is to assess the candidates (one at a time), continue research into their suitability by gathering more information about them, and clarify matters so that each one can knowledgeably bid on the job.

Prepare for your interviews by creating a file folder for each candidate and stocking it with a set of plans and specifications. Whether the plans were professionally drawn up by an architect or designer, or sketched out by you on a paper napkin, they are essential. If you're working with a design professional, he or she will have prepared plans for you, which you will then add to the contractor-candidates files. If you are not working with anyone, you will have sketched out rough plans of your proposed project and may have found an architect or draftsman to draw them for you. You can of course submit your rough sketches to the contractors, but the more professionally the plan is drawn, the easier it will be for other professionals to decipher. And by the way, make sure everyone in your family is pleased with the details of the work you are planning to do, or at least has been consulted. Family disputes after the work begins are a major source of expensive change orders. Drawings or magazine clippings that illustrate what you want the final product to look like can also be helpful. Hand each contractor his or her folder before the interview begins.

In its broadest sense, the terms "plans" and "specifications" refer to the following documents: *floor plans,* which are views of the interior layout of both the existing house and the proposed changes; *blueprints,* or working drawings, which show all aspects of the work to be done; *specifications,* which list and describe in detail the materials, appliances, fixtures, and finishes to be used in the project; *elevations,* or ground-level views, of all sides of the structure; and *sections,* which are cross-section views of certain aspects of the project. There may also be *detail drawings,* or sketches of ornamental additions, such as moldings or cornices.

A Glossary of Blueprint Labels and Symbols

Educating yourself about the different elements involved in the construction of a house will help make the building process a much smoother experience. A good place to start is by learning to read blueprints. Just a few minutes of study will help you understand how the structural elements of your home are rendered, enabling you to communicate more easily with contractors and tradespeople.

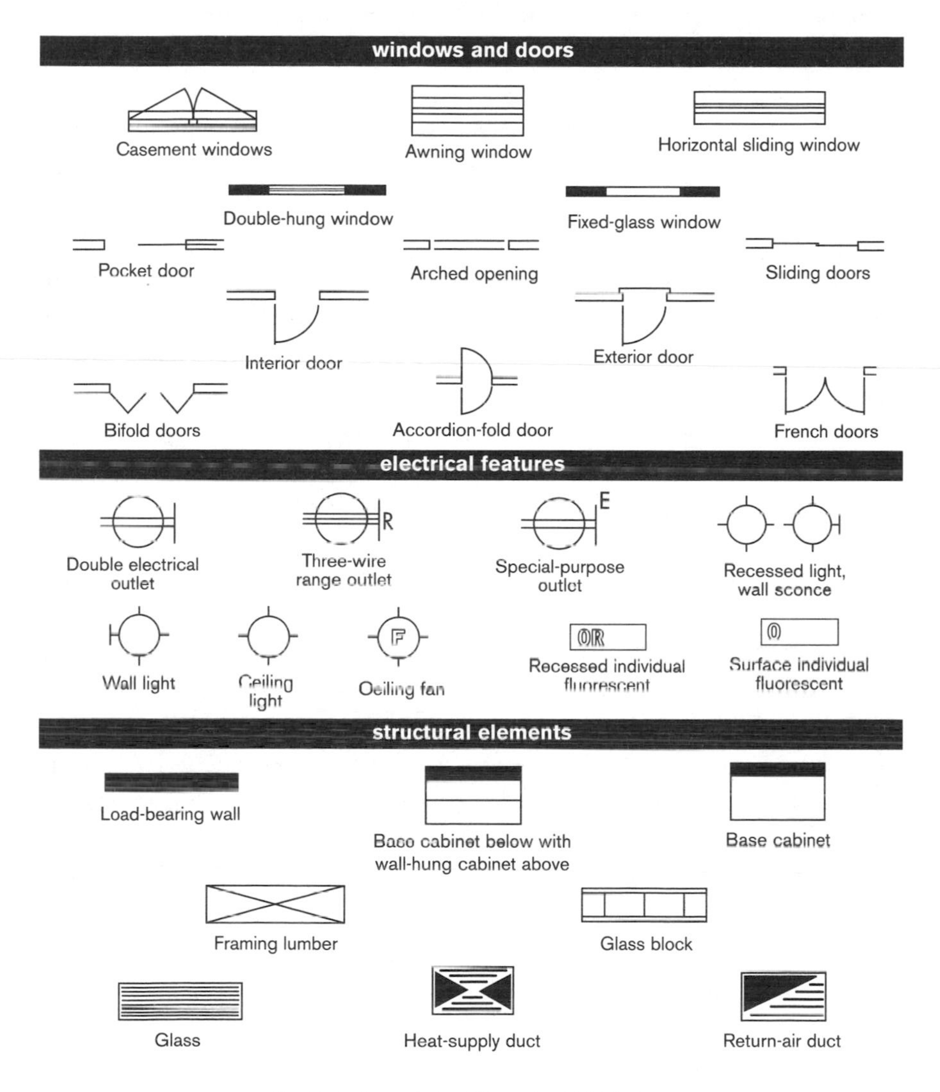

If your project is small and simple, you may not need all of these elements; floor plans and a specifications list will probably suffice. A two-story addition to your home, however, may require the full complement. In that case, you should hire an architect, if not to design and oversee the entire job, at least to produce all of these documents for you.

These documents are as important to your contractor candidates as they are to you. When you ask for bids, you will give all of this material to the candidates. Thus they will all know exactly what the project entails, and all will be able to bid on the same thing.

In his book *The Big Fix-up,* Stephen Pollan suggests living with the plans for a while before starting the bidding process. The extra time you take studying the plans and thinking about them can help minimize expensive change orders once the project is under way. "Try to envision yourself in the house after the renovations have been done," suggests Pollan. "Is there anything you'll miss. Have you forgotten anything? Pay particular attention to electrical outlets, telephone jacks, cable TV plugs, and storage space. Think about all of the things you do during the course of the day. Where will you do them? The longer you spend studying, considering, and just plain looking at your plans, the less likely you'll want expensive additions once the project has started." Other things to consider when ruminating over your plans: What's the view from your new windows? How will the sun enter the room? Is there any time of the year when there's likely to be glare when you're sitting down to meals or watching TV? Is privacy an issue?

Now is also the time to think about extra work the contractor and crew might do while your house is going to be torn up anyway. For instance, while the electrician is here, why not add some extra outlets that you'll be needing when you get computers for your kids' rooms? Or while the walls are open to receive upgraded plumbing pipes for the new kitchen sink and dishwasher, why not have rough plumbing installed for the powder room that you plan to build next year on the other side of the kitchen wall? These extras may add to the price of the job, but it makes good sense to take care of them now. And in any event, it will be a lot cheaper than adding them with a change order while the job is already in progress.

Create files for yourself too, one for each candidate. Prepare a list of questions you will ask the candidates. And of course you'll need a pad for your notes on what each interviewee has to say about the project and in answer to your questions. Pay careful attention to questions he or she asks you. It's possible that these questions may give you some good ideas that you can use, no matter whom you hire.

During the interview you should discuss the project in detail, expanding on the brief description you gave during the initial phone contact. If timing is of the essence, tell each contractor so and give him or her the completion date you must meet. If you want only top-quality materials or must have a certain brand of appliances or windows or ceramic tile, or if you insist on three coats of paint on all the walls, mention these things now. Fulfilling these requests may add time to the job and will certainly affect the bid. And if you absolutely cannot spend more than a certain amount of money, now is the time to share that information.

The more specific you are now about the project, the more accurate the bids will be. Don't just say "paint the kitchen off-white." Give details such as, the paint's brand, color, and finish, and which paints to use for walls, trim, and doors. For windows, the specs should detail the manufacturer, type, size, finish, and model number if applicable.

For an added dimension on the bids, encourage your candidates to come up with some ideas of their own on the project. Do they think it could be improved aesthetically? Could you save money by approaching it differently? Do they see any mistakes in the plans? If so, how should they be corrected? To keep the bids comparable and easy to assess, make clear that these suggestions should be listed on a separate piece of paper, not as part of the bid.

The way a contractor responds to the opportunity to make suggestions may reveal valuable information—his or her willingness to spend time with your plans and drawings, creativity at assessing a project, and savvy about cost-saving materials. This last skill is a real bonus if you're working with an architect or designer who may have specified high-cost products when a more economical ones would do just as well. An apt example given by Stephen Pollan involves an architect-designed pool house. The architect specified redwood for the little structure, which, says Pollan, "...would indeed have looked wonderful and may well have lasted for a century. But it would also have cost a small fortune." The general contractor suggested that cedar siding stained to look like redwood would cost half as much and would look as good. "Granted, it would probably last 50, not 100 years," says Pollan, "but this was a home, not a monument."

Ask that bids be broken down so that each element of the project and each subcontractor's specialty is listed separately—site prep, framing, electrical work finish carpentry, and so on. This will allow you to easily compare and contrast the bids you get, not only for total cost but also for the cost of each aspect.

And this breakdown will help you spot areas where costs may be too high. Most general contractors probably have favorite subs that they use over and over, and some of these people may be expensive. If, for instance, an electrician's bid looks high, you can always ask the contractor to get a figure from another, less pricey, sub. Experts also suggest that you ask your candidates to estimate their charges for such extras such as installing electrical outlets or adding one more coat of paint. If you request the extras before work begins, there would probably be no charge or only a minimal one. But after the project is under way, there may be a penalty charge for these requests. You'll want to negotiate this at the bidding stage.

Heads Up: Who Chooses the Materials?

You should be aware that many contractors prefer to use certain materials and specific brands of building products, and may not be thrilled about substituting materials specified by the customer. There may be good reasons for their choices. The materials may be readily available in your area or easy to work with or particularly suitable for the local climate. Or some of these items may offer a high profit margin for the contractor. According to Sven Swenson, a retired contractor from a small town in Minnesota, vinyl-clad windows are one such product. "They are marked up big and are easy to install and contractors like working with them," he says, "but there is no logical reason for a contractor to reject what you want unless it is a way-out product that's practically impossible to find."

If the contractor objects to your choices, listen to his or her reasons with an open mind, but do not allow yourself to be pressured into making any promises. Then bounce these ideas off the other contractors you are interviewing. Their responses, plus the fact that you have already done your own homework on the project, will help you make a decision.

REFERENCES

Now is also the time to ask for references of various kinds. You will want bank and business references, so that you can check each candidate's financial solvency. At this point you will also want the license and insurance information that you asked for in the initial phone call, so that you can check those too. If any candidate has forgotten to bring this material along, ask him or her to call you with it or drop it off as soon as possible; this will make it clear that without these verifications, you will not be proceeding to the next step in the process. Beware if anyone balks at producing proof of licensing and insurance.

If your calls to the Better Business Bureau or consumer affairs agencies have revealed any complaints, mention them now. You have already eliminated from the running any candidate who has had many complaints or has failed to address and resolve them. The ones you will be asking about now have probably been resolved, but even so, it's a good idea to hear the contractor's point of view on the dispute and to see how he or she reacts to being asked about it. The problem may have been exaggerated by a difficult homeowner, but if you get a harsh or vitriolic response, or one that suggests that the contractor is not adept at managing business matters, you may want to consider that a red flag.

Ask the candidates for the names and phone numbers of previous customers whose projects were similar to the one you are planning. The ideal list would include a job that is in progress and another that is two or three years old, which will enable you to see both recent craftsmanship and how the work has held up under a couple of years of use. Make it clear that your intention is not only to talk to these customers on the telephone but also to visit the projects.

RED FLAGS

During the interview, be on the lookout for responses and attitudes from your candidates that may give you a queasy feeling about a candidate. Heed those uneasy feelings, especially if they are accompanied by any of the following red flags:

- The firm or contractor you are interviewing has been in

business for less than five years. We're not talking about a going concern that has recently relocated to your area. Treat that candidate as you would any of the others. But if this is a just-formed company, you may want to move on. According to the National Association of Home Builders, nearly 90 percent of all contractors go belly up for reasons of insolvency or incompetence within the first five years.

- A salesperson from a contracting firm says that your home will be used for advertising purposes and a sign with the company's name on it will be placed on your lawn. This is fine *after* the job has been done to your satisfaction, says N'ann Harp. "Smart Consumer Services recommends against allowing a company or contractor to put yard signs up during the job—which is when they usually want to do it—because the owners don't know yet if the job will turn out satisfactorily and they will have become parties to free advertising that they may later regret.

  "By the way, this is *not* normal procedure, no matter what a contractor might say," Harp continues. "If the job goes well and the owners are happy, satisfied customers, it would be perfectly acceptable to display a sign. But the company's got to earn it." View with suspicion a contractor who tries to insist on this perk.
- The contractor seems edgy about giving you names of previous clients or tries to avoid it altogether. Contractors who do good work will be pleased to show it to you. If they make excuses, you have every right to be suspicious. Beware, too, of a contractor who tries to substitute photos of a previous job for an actual visit to the site by you.
- A contractor candidate doesn't have a license if one is required in your area. Yes, we have earlier stated that licenses do not prove ability or honesty. But failure to obtain one may indicate a sloppy attitude. In addition, uncertified, unlicensed, and uninsured home-improvement work may be illegal in your state. If something goes wrong during the job—or later, as a result of the work done—you

will have no legal recourse and may have to absorb the cost of repairing the damage.

- A contractor candidate claims he or she has already got materials and wants to pass them along to you at a discount. These materials are probably ungraded or below grade-minimums for code; or they may have "fallen off a truck." Small- to medium-size contractors rarely buy in volumes that could yield big discounts and rarely have large inventories of material. If they do have large inventories, it's probably because they badly misjudged quantities on a previous job, which doesn't speak well for their estimating abilities.
- Another red flag should wave wildly if the potential contractor seems unwilling to give you such financial information as bank or business references or gets indignant about being asked. Anyone who gets nervous about this almost certainly has something to hide.
- An interviewee is impatient and does not seem to be listening to you or exhibits a dismissive, patronizing attitude in response to your questions.
- A salesperson or contractor puts pressure on you to make a decision and sign a contract right away. Do not even consider hiring anyone who uses such scare tactics as "This is a one-time-only offer" or "This low price is in effect only today; tomorrow the discount expires." Ellis Levinson, formerly the "Consumer Guy" on ABC's *Home* show, suggests that in preparation for negotiating with tough salespeople, homeowners rent *The Tin Men* from their local video store. This entertaining and instructive movie chronicles the tricks and strategies of two slick aluminum-siding salesmen. It won't offer you practical tips on how to hire a contractor, but it will make you laugh and rev you up for your interviews.
- Missed appointments should also worry you. If a candidate doesn't show up for the first interview, don't schedule another one.

One or more of these red flags should cause you to cross the candidate off your list. You don't have to make a big deal about it. You can simply say, "Thanks for coming by."

CONTRACTORS' RED FLAGS

Contractors get nervous too. Here are some homeowner attitudes and responses that might make *them* want to walk away from a job.

- *Stinginess.* To stay in business, contractors need to do satisfactory work—and get paid. Having to struggle for payment and sometimes not getting it is a perennial risk. Just as there are contractors who do a shoddy job, there are customers who are so stingy that they try to get out of paying their fair share. One trick customers sometimes use, says Stanley Pritikin, a retired g.c. who did most of his work in New York City, is to refuse to come up with the last payment. "Homeowners think a contractor probably won't bother to sue over that last payment," says Pritikin, "so I always saw signs of stinginess as a red flag. If the first thing I heard was, 'I don't want to spend much money on this project,' I was out the door."

  Contractors are also leery of potential customers who aggressively pressure them to continue lowering the price while at the same time asking for add-ons.
- *Doubts about financial solvency.* Contractors sometimes worry about the customer's ability to come up with the money. Is their loan actually funded? How's *their* credit?
- *Reluctance to set a budget.* "If customers are honest about what they can spend, I can bid accurately," says Pritikin. Unwillingness to discuss budget issues and boundaries creates frustration and poor communication.
- *Bad vibes.* "This rapport thing works both ways," says contractor Sven Swenson. "I'm not comfortable if I sense that the homeowners don't trust me or are always watching me, thinking I'm going to cheat them." Another problem, says Swenson, is lack of respect for his skill. "If they checked me out well, then they should know I can do the job. Sure, they want to know what's going on, but I can't do my best work if the clients are always questioning me." Jim DeWitt, who has been a general contractor and builder in Delhi, New York, for 35 years, agrees that trust is important. "I watch people in the initial interview," he

says. "If their eyes are darting around and not meeting my eyes, something is wrong. I get a feeling of tension and lack of trust, which is not good. For this kind of relationship you need to have trust."

N'ann Harp cites some other attitudes that make contractors nervous and foster bad vibes: rudeness or a condescending know-it-all demeanor; flirtiness; hyper-anxiety. "Contractors also want to run from neatniks who may have unrealistic expectations about the mess of remodeling," she says.

- *Time wasters*. Just as some people go window shopping for houses and waste the time of real-estate agents, some homeowners interview contractors just for fun and for ideas about something they may, but probably won't, do in the future.

A Quick Look at What You Need to Know

- Registration and policy numbers for contractor's license and insurance coverage
- Bank reference with a note that gives bank personnel permission to divulge a financial picture
- Two or three business references
- Names and telephone numbers of two or three previous customers with projects similar to yours
- A price for the job
- Bids for each of the trades involved
- A list of the costs of potential extras such as adding electrical outlets

AFTER THE INTERVIEWS

No, you're not finished yet. You'll need to call the bank and business references the contractors have given you. Explain your situation and say that you want to know if the contractors in question are financially solvent. "If you approach the matter without too much undue formality, you'll be surprised at how forthcoming people will be with information," says Stephen Pollan. Some bank officials, however, cannot give out this kind of confidential information without a written OK from the contractors. Urge all your candidates to give you this permission and view with alarm those that resist. Here's why: If the person you hire goes under in the middle of your job, you're probably not going to recover any money. In fact, you'll have to lay out more to get someone else to finish the job. A detailed, solid contract, which we discuss in the next chapter, will protect you against total disaster should this happen; but even so, you want to avoid it if at all possible. Do everything you can to ascertain that your candidates are solvent and likely to stay that way.

In addition, you must call the appropriate agencies to verify that licenses and insurance coverage are in effect and when they expire. Contractors have been known to cancel their insurance as soon as they think they've got the job. And a license may have been revoked or suspended, a fact that you would not discover without making that call.

THE INSURANCE ISSUE

While you are at it, talk to your own insurance agent. Describe the project you are contemplating and ask whether or not you are sufficiently covered for the situation by your existing homeowners policy. If your remodeling project will increase the value of your house, you may need to update your homeowners insurance. Your agent may suggest that you insure the house itself for "replacement value," that is, for at least 80 percent of the actual cost of rebuilding, not for the location-sensitive market price. Possessions can be covered for "actual cash value," the original price minus depreciation. But because this figure often falls short of current costs, experts recommend paying 10 to 18 percent more for a replacement value policy for your possessions as well. Your agent can supply instructions and forms for compiling a comprehensive household inventory.

Your home security may be compromised during construction,

especially if exterior walls will be torn down. Before you begin, you should also review your homeowners policy for weather damage, theft, and fire. Most policies will cover building materials only after they are installed. So be sure that your contractor is covered with an installation floater for all products and materials until they are installed. In addition, you should be insured for the cost of the construction before any renovations begin.

Insurance agent Keith Slater points out that with existing homeowners policies, most people are covered up to the limit of the policy only, a figure that may be inadequate for remodeling protection. "When you're remodeling, you need to call your agent and increase the value of the policy to include the value of the remodeling project. If you added a room for $60,000, you would probably want to increase your policy by that amount." Let your agent guide you in the particulars of your situation, says Slater.

Why such an emphasis on insurance? Because construction sites are chock-full of pitfalls and opportunities for injury and property damage. Power tools and ladders can pose a danger, construction debris makes walking around the site hazardous, wheelbarrows and other intriguing pieces of equipment can entice children to play. Unless you—and, equally important, your contractor—have the proper coverage, you could be heading for big trouble. Your updated homeowners policy with general liability should cover most of your needs, including injury and damage to property.

If both you and your contractor are adequately covered, you don't need to worry about facing financial ruin when you encounter problems like these:

- One of the subs backs up a truck in your driveway and hits a neighbor who has come to visit.
- You trip over a short stack of lumber and badly sprain your ankle.
- A contractor's assistant falls off a scaffolding and is seriously injured.
- Concrete spills out of a mixer and oozes into landscaping that your neighbor has just invested a lot of money in.
- A broken pipe spurts water all over your new living room carpet.

The Insurance Risk: A Cautionary Tale

Just in case you're tempted to skip this annoying insurance probe, here's a glimpse from N'ann Harp at what might happen if you hire an uninsured person.

"Entry into the danger zone of unlicensed-and-underinsured-contractor territory usually happens with a person who's been doing odd jobs for you for years, a person you like and trust. One day you ask him to replace some switch plates; and, oh yes, while he's at it would he mind putting in a dimmer switch in the hallway? You know he's not a licensed, insured contractor. He knows he's not a licensed, insured contractor. But you like him, and this little job will give him another couple of billable hours. In actuality you are putting him in the position of having to refuse your request, or unwittingly exposing himself (and you) to a serious hazard lurking in your wall. Say that your house is old and the wiring is faulty. Or maybe your house is new, but the electrical inspector had a bad hangover that day and never noticed the uncapped 220-volt wires left exposed in the wall, intended for an appliance that has yet to be installed.

"You run some errands while your handyman does the work. When you return, there are fire trucks and an ambulance in front of your house. The unconscious, badly burned man, who had a heart condition, is being taken to the hospital, and your house is in flames.

"You can't imagine what your insurance agent will say when she learns that the man doing electrical repairs at your house was unlicensed and uninsured. The man may require heart surgery and skin grafts. He won't be able to support his family. Who is responsible?

"Murphy's Law, many believe, was created expressly for the home-improvement industry. Many have tried to beat the odds of something hideous like this happening, and many have failed. Think very hard before putting someone in a position like this. You might argue that the handyman should have refused to do the work and is therefore responsible for the outcome. You might cling to the belief that your homeowners policy covers this kind of injury. A personal-injury attorney, however, would drool over the chance to take on a case like this and nail everybody.

"Find out from your state consumer-affairs office what guidelines exist for hiring an unlicensed worker and what legal risks you accept by doing so," concludes Harp. "And resist the impulse to give a trusted handyman a few hours work that requires him to perform licensed contractor work. It's not worth the risk."

But contractors should be covered by more than just general liability. Small- and medium-size contractors are often underinsured, so check your candidates carefully.

Workers' compensation pays for medical expenses, lost wages, and rehabilitation for any employee injured on the job, regardless of fault. In most areas of the country, this sort of coverage is mandatory, but in some states a contractor with fewer than five employees may be exempt. **Contractor's general liability** covers damage to you, your family, anyone who happens to be on your property, and your property itself. It also covers such mishaps as windows broken during installation, floors that warp after work has been completed, even faulty materials. The contractor is not covered, however, for faulty workmanship. If the general liability policy does not include **vehicle insurance** that provides coverage against all possible claims involving the use of company-owned cars, trucks, and so forth used in construction work, the contractor should get an amendment for that purpose. Other types of insurance that are sometimes recommended by industry experts—**product liability; completed operations liability,** which provides coverage for physical injury *after* the project has been completed but resulting from the work performed; or **non-owned vehicle liability,** which covers employee-owned vehicles being used for work purposes. If you will be doing part of the work yourself, ask your agent if special coverage is required in case you are injured.

Ask to see your candidates' certificates of insurance, which will show the name of the carrier, the expiration date, and the policy limits, and ask to be named an additional insured on the general liability policy. Look carefully at the policy limits, says Keith Slater. For a major remodeling project, they should be about $500,000. "Less than that is not adequate to cover the kinds of problems that could come up," he says. "To increase coverage to that level is not expensive, even for a small, one- or two-person operation—only about $50 a year." Consider asking the contractor to up the policy limit if you feel it's necessary. For more data on insurance issues you can call the Insurance Information Institute help line or visit their web site. (For details, see Chapter 7.)

INTERVIEWING PREVIOUS CUSTOMERS

Call each reference the contractor has given you and set up appointments to talk to them. This part of the decision-making process is time-consuming, but it is an important and effective way to get an accurate sense of a contractor's work and his or her relationship with other clients. After your initial discomfort wears off, it's easy. Other homeowners actually like sharing their experiences with you. And they can tell you exactly what you want to know about the contractor in a way that no referral service or trade association can.

Your visits don't have to be long. You can easily find out what you need to know in 15 minutes or so. Take a notepad with you so that you can record the homeowners' answers to your questions. Don't worry if they think you're obsessive. They very well may think so, but chances are you'll never see them again. And besides you can't hope to recall everything they tell you if you don't write it down.

Ask objective questions. Don't focus on how much the homeowners liked, or did not like, the contractor's personality. Instead, concentrate on their assessment of his or her skill, work habits, and reliability. Address the issues that matter most to you. Some of your questions might be:

- What type of work did the contractor do for you?
- Did the project start on schedule?
- Did you have any difficulties with the local building authority?
- Did the project go as planned?
- Did the contractor come to the site daily?
- Was the g.c. easy to work with on a daily basis?
- Did the g.c. give you regular progress reports?
- Were you comfortable asking questions about the job?
- Did you make any changes? If so, did that process go smoothly?
- Were there any problems? If so, what were they?
- Was the crew punctual and did they stay until quitting time?
- Did the crew clean up daily?
- Did the crew take too many breaks, or excessively long ones?
- Were there delays? Why?
- Was the job completed on schedule and within the budget? (Most jobs aren't, so don't let a "no" answer put you

off. Do, however, try to find out why. Were delays caused by on-site surprises such as rotting subfloors, a beam that didn't show up in the plans, or a supplier's strike? Or was it poor management by the contractor?)

- Were you satisfied with the work? If not, what in particular dissatisfied you?
- Did you have any problems after the job was completed? If so, was the contractor willing to resolve them?
- How did the work hold up?
- Would you hire this person again?
- Would you recommend this person to your friends?

ASSESSING PREVIOUS JOBS

Visiting jobs in progress gives you a chance to see the work as it is being done; reviewing completed jobs reveals the quality and durability of the work. But of course you need to know what to look for.

When you're examining a job in progress, study the framing. First, walls should be framed with at least 2 x 4 lumber in areas that experience severe cold weather, 2 x 6 lumber is preferable because it allows the builder to pack in extra insulation. There should be no more than 16 inches of space from one wall stud to the other. Floor joists should be sized for the drywall span, also no more than 16 inches apart. Subflooring should be 3/4-inch plywood. With drywall construction, the spaces between panels of the wallboard should be taped and filled in with joint compound so that the walls look seamless with sharp corners and no dimples.

When you're looking over completed jobs, walk slowly through the rooms that were remodeled and look closely at details; a careful look will usually reveal sloppy work. Study the finish carpentry. Look for neat cuts and tight corners. Moldings should be nailed tight to the walls with no gaps showing. Look askance at cabinets that are nailed to the walls; they should be screwed in. Scrutinize paint jobs, looking for neat trim, even coats with no sign of brush marks, sharp lines where wall paint meets ceiling paint. Pay attention to floors, whether vinyl, ceramic tile, or wood. On vinyl flooring, look for a centrally located pattern, equal-size borders on all sides, no spaces between the floor and the bottom of cabinets. With ceramic-tile floors you want to see a level surface, even spacing between tiles, unobtrusive grout lines. When you're assessing wood floors, beware of visible nails, signs of cupping, gaps between boards.

Does all of this checking and cross-checking and double-checking seem tedious and unnecessary? It probably does, but before you decide to skip it, listen to what N'ann Harp has to say about a homeowner's worst nightmare:

"If you have hired a contractor without checking references thoroughly beforehand or enlisting a lawyer to review the contract before you sign it, you are your own worst enemy. You are setting yourself up for the homeowner's worst nightmare. It goes something like this. Your home-improvement project, which started happily enough, gets interrupted for one reason or another. Then the problems really start to snowball. You argue with the contractor and your project is left incomplete. A complaint is issued or a lawsuit is filed. You don't get any money back but you have to pay your lawyer's fees anyway. You hire a new contractor to finish the job, and to add insult to injury, you pay for the same work all over again plus your legal fees. You are emotionally frazzled for months. In the end, you paid three times as much as you had budgeted for the project. And as a bonus, you now actively distrust all home-improvement contractors."

If you really hate the idea of going through the reference-checking process—or you know that you're too busy to do a proper check yourself—Harp's organization, Smart Consumer Services, will do it for you. For a fee (at this writing it was about 50 dollars), SCS checks commercial, insurance, license, and past customer references of your candidate and gets reports on him or her from the Better Business Bureau and consumer-protection agencies. (see Chapter 7 for details.)

BIDS

By now you have eliminated several candidates for one reason or another. The rest of them will be bidding on your job. The playing field is pretty even at this point, and it is via the bids that you'll arrive at your final decision. Like all aspects of this important endeavor, the bid-assessing process could take a little time. Don't rush it.

Establish a deadline for bids. Estimating is a time-consuming process for a busy contractor, so you may not get the estimates exactly when you'd like. But naming a definite date will produce better results than being vague about it.

As you checked references, quizzed various people about the contractors on your list, and talked to previous customers, you probably took some notes. Put them in the file folder you have set up for each candidate. By the time you have finished checking and filing the references, the candidates will be calling, ready to submit their bids. Most will want to drop the bids off in person. That's fine; it gives them a chance to explain to you how they came up with their figure, and it may be helpful to you to hear their reasons. They may also use this face-to-face meeting to try to sell you or to pressure you into making a decision. Resist any such pressure. Listen to what they have to say, take notes if you find it helpful, thank them for their efforts, and say you'll get back to them soon. Don't let anyone pressure you into making a decision then and there.

An Estimate is Not a Bid

By the way, it's important to know that there is a difference between an estimate and a bid. The former is an estimated price for the work, a rough, ballpark figure that the contractor may mention when he or she first looks over the job. It is a rather informal stab at what the job may cost, based on vague documents or verbal descriptions and may help you decide whether you want to proceed with the project or not; but it is not to be considered formal and binding. Many contractors don't estimate the job at all, but wait until they have worked up an actual bid. Do not accept a proposal headed by the word *estimate*, even if it contains a description of the work and a price. Such a proposal may not be legally binding.

A bid, on the other hand, is a formal, legally binding statement that a contractor will complete a job, described in detail in the document. "A bid states that the contractor intends to do the work for that amount of money in that amount of time," says Oakland, California, attorney Ann Rankin.

In order to prepare a bid, the contractor must consider the entire job and figure in all costs from beginning to end, plus a reasonable profit. Some projects, however, are quicker and easier to bid than others. If the project is a simple and straightforward one, such as the installation of siding or roofing, which are measured in terms of squares (100-square-foot units), the contractor may be able to eyeball it and give you a price right there. (Fine; but ask for this kind of bid in writing as well.)

But any job that modifies the existing structure of the house, inside or out, such as adding a wing, refiguring interior space, or installing a new kitchen, will be more difficult to assess. For one thing, there are many more elements to consider. For another, the contractor will have to get figures from a group of workers (contractors and laborers), obtain costs for various kinds of materials and appliances, add something for overhead and profit, and blend it all together into a total price for the job. Some contractors, correctly guessing that others are also bidding on this job and wanting to be competitive, may adjust their own final figure somewhat. The more someone needs the job, the more willing he or she will be to cut profit margins.

Get the bids in writing. They may be written on a standard form or a company letterhead, or they may be on a plain sheet of paper. But "in writing" is the key. Under no circumstances should you accept an oral bid, either in person or over the phone. If a candidate wants to give you a preliminary figure orally, that's OK. But do not accept a bid as final unless it is in written form. In addition, it should be signed and dated by the contractor. Not, signed, by the way, by you. Be careful about what you sign. A bid or estimate that you have signed could be construed as a contractual agreement, one that you may not be ready to make.

Give yourself time to read and decipher all bids carefully. Be sure they are comparable, that the contractors did in fact bid on your specifications. Does each bid include the brand of appliances you specified, the type of paint, the kind of cabinets? It will be very difficult to compare bids unless all prices are based on the same information.

Trimming costs

If the bids shock you—and they probably will—talk to your contractor candidates about trimming costs. Make some of the following suggestions and see what they think. If any candidate is reluctant to discuss cost-cutting strategies with you, cross him or her off the list.

- Rethink. Consider reducing the size of the addition you plan to build or making individual rooms smaller or even eliminating a room altogether.
- Postpone. If your project is a two-story addition with a family room on the first floor and a master suite on the second, you could leave the second floor unfinished and concentrate on the family room for now. The second floor will be framed and roughed in; when you feel you can afford it, you merely do the finish work. Doing the work in phases will probably not bring down the overall cost of the project; but this alternative to laying out all the money at once will make it easier on your pocketbook.
- Economize. Sometimes homeowners can bring costs down by choosing more economical materials and finishes. Opt for laminate instead of granite for your kitchen counters, standard bath fixtures instead of top of the line, a vinyl floor instead of ceramic tile. You might also save by buying construction materials from clearance, carload, and suppliers' distress sales at lumberyards or home centers, but check quality carefully before you buy. Other possibilities: suppliers' seconds or buying directly from the manufacturer. Ask your candidates if there are any materials they can get at especially low prices.
- Do some of the work yourself. A good idea for a talented and experienced do-it-yourselfer, but for anyone else it's risky. See Chapter 1 for details.

  However, you could easily save a little money by taking on very simple tasks such as daily cleanup; if you're saddled with a really tight budget, why pay for someone to sweep and straighten up the job site when you can easily do it yourself?
- Negotiate. You're in a position of power. Your candidates know they have competition; they all want the job. It's possible that one or more of them will reduce their bid, especially if they know someone else's came in lower. You might get the ball rolling by letting the contractors know you're willing to adjust your schedule to theirs. A g.c. who's busy now but has no jobs planned for a few months from now may be willing to lower the bid in return for a sure job.

What if a contractor submits a bid that does not carefully follow the specifications you spelled out? If all other things about him or her seemed positive, you might want to call and ask why. The contractor may have prepared the bid too hastily or misinterpreted the plans, in which case you could ask for a new bid. However, if it seems to you that the discrepancy comes from a lack of knowledge or experience, cross this person off your list.

If you have been clear and specific about the project in your initial interview with the candidates, the bids you receive will be "apples to apples" and not "apples to oranges." If you were vague, you'll get bids that are difficult to compare.

Bid Basics

- Insist on receiving all bids in writing. Never accept a verbal bid. Don't accept bids over the phone, even from someone who has seen the job. If a contractor is not willing to spend the necessary time to study a job and then bid on it properly, how careful will he or she be with the job itself?
- Make sure that the contractor who will actually do the work looks at the site and personally bids the job.
- Be sure each bid specifies what materials the contractor is to use.
- If a bid seems too good to be true, it probably is.

You may be surprised at how different the bids are from one another and perhaps from what you expected. Don't be concerned about a high-medium-low range, as long as all of them are in the same ballpark you're playing in. But if all the bids are higher by, say, 15 or 20 percent than the figure you expected (or can afford) or if they differ so widely from each other that it's hard to believe all of the candidates are bidding on the same job, you may have a problem.

The problem could be with the plans. Did you do your homework? If all the bids were high, you probably miscalculated what the job would cost. If the bids differ greatly from each other, your plans may not have been specific enough, leaving too many decisions about materials up to the contractor. Were you working with a design professional? He or she may have failed to keep the plan in line with the cost level you requested.

In either case, you will have to go back to the drawing board and refigure the job with a design that's within your budget. Call your contractor candidates and tell them you will be preparing a fresh set of plans for them to bid on and will have them within a week. You may lose a candidate, but most of them will be willing to wait. If you're working without an architect or designer, you might get back on budget by asking each candidate to suggest some ways that you can trim costs.

If the fault lies with your architect or other design professional, a few tweaks of the design may fix the problem. If the changes required are major, it could cost you more money, which is annoying, but you should probably resist the impulse to fire the architect at this point, unless you're willing to take the time to hire another one and set the project back several months.

Compare the bids closely, looking for discrepancies. Does the allowance for kitchen cabinets differ greatly in one bid? Why? Were you not specific enough about the kind of cabinets you wanted, or does one of the candidates have a source for quality cabinets that is less expensive than the ones you specified? Alternatively, a candidate may have suggested a costlier brand that he or she feels is superior to your choice. Perhaps the price for the kitchen floor is considerably lower in one bid than in the others because one of your bidders doesn't think you need a new 3/4-inch plywood subfloor in that space. "The new floor can go in right on top of the old one," he might say, shaving $1,200 off the price. If this kind of thing happens, keep an open mind. Experienced, skilled contractors have seen many projects like yours and have developed many effective, creative ways of dealing with them, not to mention the wealth of valuable sources and inside dope they are privy to. Maybe the bids for painting differ, not because you failed to be very precise about the brand you wanted but because one of your candidates plans to do the painting himself, saving you—and him—the cost of a hiring a painting subcontractor.

Take a look at the prices for other subs. They should be reasonably close in all the bids. If one contractor failed to include a certain trade, or conversely, if only one candidate included a trade you thought was necessary—make a note to ask about it later. It may be that he or she made a mistake or noticed something the other candidates didn't see or has sneaked an unnecessary expense onto your project. If you notice that subcontractor prices in one bid are out of whack with prices for the same work in the other bids, ask about that too.

When you ask about these discrepancies, you may get a logical response that makes sense to you. If not, ask for a new price on that item or cross that candidate off your list.

Time and Materials Bids

Some people view time-and-materials deals with horror; others think they are a good idea in certain circumstances.

In this kind of arrangement, the homeowner is billed for goods and services, in addition to an agreed-upon profit for the contractor. Time-and-materials bids are often used when projects are difficult to estimate, if, say, plans are vague or the house is in such bad shape that the contractor can't predict what problems will arise from one day to the next. In these cases, a time-and-materials bid could save you money. Other contractors, who may not want to deal with vague plans or really rundown houses unless they can make a lot of money at it, may submit sky-high bids to cover all eventualities. With a time-and-materials, pay-as-you-go situation, you can circumvent extra-high bids and deal with problems as they arise and as you can afford it.

Don't enter into this type of arrangement, whatever shape your house is in, unless you have thoroughly checked out the contractor and feel sure that he or she is trustworthy. Also, a time-and-materials deal is probably not for you unless you are in a position to drop in on the job often and satisfy yourself that the time you're being billed for is accurate and the materials you're paying for are for your project only, not for other jobs the contractor may be working on.

Contractor Sven Swenson used the time-and-materials setup for many of his jobs. "It was advantageous for me as a contractor and for my customer too. With full disclosure of materials costs and an honest accounting of my time, I could lock in a reasonable profit and probably save my customer some money. If I had to submit bids the conventional way, I would have had to factor in what I perceived might be problems and then add a cushion to protect myself."

High vs. Low

If your specifications were precise, and if the contractor candidates followed them carefully, the bids you receive should not vary greatly. But do be prepared for some diversity. For a roofing and siding project, a woman we know was stunned to receive a high bid of $17,000 and a low one of $7,500. Another homeowner, who was planning a two-story addition to her suburban house, received bids that varied by $20,000.

There is much analysis in the home-improvement field about which bid is best—high, middle, or low. Conventional wisdom states that middle to upper-middle bids are best because they allow for labor and materials cost increases and other unforeseen situations, and they provide some insurance that the contractor will not have to cut corners to finish the job.

Conventional wisdom also suggests that you view very low bids with suspicion. Some contractors may be so desperate for a job that they put forth a rock-bottom bid, hoping to appeal to the budget-minded homeowner. Then, because they can't actually do the job for this low figure, they cut corners or keep upping the price as the job goes on. Low bids may also indicate low wages for subs, which could easily produce shoddy work, or overlooked items that will have to be added through change orders.

THE FINAL DECISION

By now you have found out everything you need to know about your candidates, including what they will charge you to do the job. You are now equipped to make an informed, intelligent choice.

Price and competence should be your topmost concerns. Or, as Stephen Pollan puts it, "You need to balance and weight four factors: experience, motivation, skill, and cost." Of course, you'll want the lowest price possible for the best job possible, but you will not necessarily want to take the lowest bid. You may decide to go with the bid in the middle, a price that isn't outrageously high and yet is not so low that you're suspicious about it. Committing to an additional 5 or 10 percent over the lowest bid is money well spent, provided the contractor in question does good work, has been checked out as trustworthy, and seems to be cooperative, flexible, and easy to get along with.

Because you have thoroughly investigated all your candidates, at this stage in the process you can be reasonably confident that they will do a good and honest job for you. All these things being equal, you'll want to consider the sincerity and enthusiasm of the customer references you solicited and the way you feel about spending weeks if not months with this person. You should also take a look at your conversations with the candidate. Did you like his or her personality, feel a sense of trust, detect such desirable qualities as patience, an even temper, and a sense of humor?

Once you have made your decision, be considerate and let the candidates you did *not* select know right away They may be holding off taking other jobs until they hear from you.

Should your architect choose the contractor?

If you are working with an architect or interior designer, you can elect to be less involved in the selection and bidding process, although you must not abdicate control over the contractor decision completely. "If the clients agree, we help choose the contractor and also handle the bidding process," says architect Steven House of House + House in San Francisco. "It makes sense for us to do this because we have a track record with certain builders and contractors and have had experience with a number of people; the homeowners probably haven't. We can act like a clearinghouse, recommending people that we know have integrity and skill. We'll suggest two or three candidates and get bids from them on our plans. Then we'll help the clients make the decision." Or House + House may negotiate a contract by bringing the contractor in early so that he or she can help develop the budget as the plans evolve. With this method, the contractor becomes, in effect, a cost estimator as well.

Although New York City designer Rick Shaver prefers working with one of the two contractors he's worked with for years, he has no objection if clients want to bring other candidates in. "But if they do," he says, "I like to be in on the interviewing and the bid comparisons, and to check out two or three of the other contractors' jobs myself. With those safeguards built in, whatever person they hire usually works out fine."

There are many advantages to choosing a contractor with whom your architect has established a good working relationship, but there's also a downside. You will be relying on the architect to supervise the general contractor and the quality of the work, and you don't want them to be such good buddies that the supervision amounts to nothing more than a rubber stamp.

And if you will have both an architect and a contractor on the job, you must also consider the antipathy that often exists between the two professions. Some architects think contractors are good for nothing but swinging a hammer and can't be trusted to read plans and execute complex designs. And some contractors see architects as trouble-making prima donnas who wouldn't know which end of a hammer to pick up. Going with a contractor your architect has worked with before, or asking all prospective contractors how they feel about working with architects, would eliminate this potential friction.

Consider one bid that comes from the contractor that your architect recommends, says N'ann Harp, but don't let that bid be the only one. "Even if the homeowners are about 95 percent certain they'll go with the construction company the architect or designer prefers, it'll help keep everybody honest if a second and third bid are also solicited.

"Trust is not the same as abdication of power," Harp advises, "and homeowners should never forget that the temptation to abuse trust is a fact of business."

four

Ironclad Contracts

There's finally light at the end of the tunnel! Using the process that we have outlined, you have chosen a contractor, you feel confident about your decision, and you believe that you can accomplish the renovation of your home in keeping with the budget you had in mind. You're all set to go, right? Not quite.

Now comes another important part of the enterprise—perhaps *the* most important part—creating the contract. We use the word *creating* advisedly. The contract you ultimately sign should be tailored specifically to your project; it should fit your needs and your goals for the renovation like a glove.

A contract is essential for both parties because both are at risk. In spite of the protection a diligent contractor search affords, homeowners engaged in a renovation face some frightening possibilities. For example, the contractor could leave town right after the kitchen is gutted; one of the subs could run over the neighbor's dog while backing out of your driveway; a violent rainstorm could roar in without warning on the day your roof is removed.

The contractor's risks are just as perilous—a customer can refuse to pay, creating a serious problem for a small business that exists from job to job. Or an obstreperous homeowner can make so many unreasonable demands and request so many complex changes that the professional actually loses money, a circumstance that happens more frequently than you might think. In fact, says N'ann Harp, "A faulty contract is the source of all those horror stories you've heard about construction catastrophes and conflicts between contractors and homeowners."

And yet, says Bruce Hahn, president of the American Homeowner's Foundation, "All too many deals are done on a handshake or a one-page bid sheet, which is not much better. Most of the complaints we get could have been avoided if both parties had used a balanced, comprehensive, plain-English contract."

The first step toward a balanced contract is to go over the bid with your contractor. His or her copy should of course match your copy in every detail. If it doesn't, correct any discrepancies you find right away. If you have changed your mind about anything since the contractor's initial inspection, discuss it now. Failing to do this now could create unnecessary complications later.

At this stage of contract negotiations, your concern should be to get everything down on paper. Yes, you did a great job by including many of the details when you put the project out for bid. All those details—materials, brand names, sizes, and so forth—will become part of the contract. And there will be more, because a contract tailor-made to your project will be very specific. By the way, do not agree to use the bidding form as a contract. Many home-improvement contractors like to do business this way, but the resulting document is too vague to give you any solid protection should problems arise.

Here's an example of how specific you should be about something as simple as a new wall. Will it be insulated? What brand? What type of insulation—blanket, rigid board, blown-in? What brand? What R-value? What will cover the interior wall. Wood paneling? If so, what brand, style, and color? Drywall? If so, what thickness? What brand and color paint will cover the drywall? How many coats do you want?

If a new or renovated bathroom is part of your project, you will want to include specifics for all of the following in your contract: flooring, wall covering, plumbing fixtures, faucets, vanity cabinet, medicine cabinet, ventilation, light fixtures, electrical outlets, switch plates, towel bars, toilet paper holder.

When you meet with the contractor to discuss all this in detail, you may have to do some renegotiating about price. For example, for the bathroom you may have initially specified a sheet-vinyl floor, then decided you like ceramic tile better as a floor covering. Unless you go shopping for the tile yourself, you will have to budget more money for this upgrade. And even if you do buy the tile, installing it will be more labor-intensive—and thus more expensive—than the vinyl flooring.

If costs seem to be skyrocketing over and above the original bid during these early contract negotiations, there are a couple of solutions. One, change your mind and go back to the original, less expensive choice, say, the vinyl flooring. Two, try to negotiate further. Using the bathroom floor again as an example, tell the contractor that you are willing to shop around, find the tile you like, buy it, and haul it home, saving him or her the trouble. In return, would he or she be willing to install it for the same price as installing the vinyl?

This sort of negotiation could work to your advantage because you are holding the money and therefore the power. The contract is not yet signed, and even if it were, you have three days to cancel before it is binding. The contractor knows that you interviewed more than one person, which may make him or her more amenable to being flexible. You are holding most of the cards, but you do not want to beat anyone over the head with them. Be friendly and cooperative, and do not use any threats, direct or implied. Contract negotiation is nerve-racking but necessary. Try to stay serene while you quietly and calmly work toward getting the provisions and protection you need. If you create an adversarial feeling now, it could cast a pall over this important relationship for the rest of the project and make the whole operation unnecessarily tense and difficult.

ELEMENTS OF AN IRONCLAD CONTRACT

To assure success, you, your contractor, and your lawyer will fine-tune the contract so that it meets your needs and protects your interests as well as possible. Here are some items you might want to include:

- Date of the agreement.
- Homeowner's name and address, and all phone numbers in case of an emergency.

- Contractor's name, business name, business address, fax number, and day, evening, and beeper numbers.
- Contractor's license and/or registration numbers.
- Architect or designer's business name, address, and phone number.
- Name of the person who will supervise the job on a daily basis, probably the general contractor.
- Location of the job site, if different from homeowner's address.
- Description of the work. If drawings and plans will be included (see next item), the description of the work can be short and simple. Otherwise, it should describe the project in detail, and to avoid making the contract unwieldy, it can be included as an appendix.
- Copies of all drawings, plans, and specifications. These items can also be made part of the appendix. Study all plans carefully before you approve them; and make sure the contract stipulates that you must sign off on the plans before work begins.
- Model numbers and/or detailed descriptions of all materials and equipment. All items that have been specified—such as appliances, fixtures, fittings, insulation, drywall, and the like—should be described in detail. And you should specify all details that you care about, down to electrical switch plates and cabinet hardware. Otherwise, items that the contractor selected, and they may be of poor quality or unnecessarily expensive. (Many home-remodeling pundits suggest "or equivalent" as appropriate language following a specific model number in this kind of clause. This is a risky policy, however, because it gives your contractor permission to determine what an equivalent is and to perhaps select the second choice instead of making efforts to find the first. Instead, insert a substitution of materials clause, which will state that your contractor may not substitute any materials or appliances without your written consent.) Here is a good place to specify which items will be supplied by the homeowner and which will be purchased by the contractor. A list of

About the Payment Schedule

The contractor will want a deposit before work starts, usually on the signing of the contract. There is no reason to be suspicious about this. He or she must buy supplies and materials in order to begin the job; and if the firm is small or medium-size, there may not be enough in the till to pay for them. This initial deposit might range anywhere from 5 to 20 percent. Or you may make a small down payment on signing the contract and a somewhat larger payment, say 15 percent, the day work starts. *Do* be suspicious if your contractor asks for appreciably more than that up-front, no matter what the reason. In a worst-case scenario, an unethical contractor will take the money and run. In a less sleazy but nonetheless undesirable situation, the inefficient contractor will use the money to pay off debts from the last job and then will have to ask you for more money very soon, a process that could continue throughout the project if you allow it. Don't allow it; you will be the loser if you do.

Subsequent payments can be made at specified intervals or as the contractor requests them. On small jobs payment is frequently divided into thirds: first third when work begins, second third when work is about halfway done, final third on full completion. Another method, often employed on larger jobs, is to tie payments to stages of completed work: a certain amount or percentage upon completion of demolition and rough carpentry, another installment after rough-in of plumbing and wiring, another after fixture and appliance installation, and so on, leaving anywhere from 10 to 20 percent at the end, to be withheld until every detail is completed. Provided you don't let the contractor get too far ahead of you, you can't get burned too badly with this arrangement, no matter what happens. It's also a good idea to hold back final payment for several weeks to a month or more to confirm that everything, such as a heating and cooling system or a new roof, is in working order. Home-improvement laws in some states regulate the amount of the deposit and the payment schedule.

Home-improvement writer Tom Philbin believes in giving the contractor nothing up front. In his 1991 book *How to Hire a Home Improvement Contractor without*

model numbers and descriptions of materials is likely to be lengthy; to avoid an unwieldy contract, put it in the appendix.

- List of subcontractors. Include the names of plumbers, electricians, roofers, painters, and any other subcontractors

Getting Chiseled, Philbin questions the one-third–up-front payment schedules recommended by many people, including, he says, some consumer-affairs departments. One-third, Philbin points out, can be a lot of money. On a $45,000 job, it's $15,000, which the customer is blithely shelling out for, well, nothing.

The danger is that with this much money in hand an unethical or financially stretched contractor may never start the job or may start and leave. Maybe $5,000 worth of work gets done; but if you've coughed up $15,000, you're out $10,000.

You can sue, of course, and with your ironclad contract, you will probably win; but that will not necessarily get your money back. You then face the problem of collecting. The solution? "Never let the contractor get ahead of you," says Philbin. "On a big job, many payments are better than chunks. For a $60,000 room addition, you could pay a few hundred dollars up front, then $6,000 upon completion of each phase of the construction—say after foundation, framing, plumbing, electrical…." You may also want to consider holding back 10 percent of each payment until you're completely satisfied with the work.

If your project is being financed by a bank or other lender, confirm in writing that payments will be made directly to you and not to the contractor.

You might also think about using a funding-control (or joint control) company to oversee the payments. For a fee, these companies hold the money for the project in an account for you and issue payments directly to the suppliers and the contractor. Contractors, by the way, generally don't like this way of doing business. With its voucher system, which gets a little complicated, it's more work for them. But some remodeling-industry experts recommend the funding-control approach for any large job, say, $50,000 or more. The fee you pay, usually 3 or 4 percent, may be worth it if it eases the pressure of monitoring the job and worrying about mechanic's liens.

To find a funding-control company in your area, ask your bank, state banking authority, or chamber of commerce for names. And check out all contenders with the Better Business Bureau before you sign on.

scheduled to work on the project, along with addresses, phone numbers, and any applicable insurance certificates or license numbers.

- List of suppliers. A list of the names, business addresses, and phone numbers of all suppliers who will be sending materials to the site will be useful for determining who will be issuing lien releases.

- Total contract amount, or final price of the job.
- Payment schedule. A breakdown of the total amount due, stating the size and due date for each payment. (See box on page 102.)
- Financing contingency. If at contract-signing time you have not received financing for the project, insert a clause that the contract is void if you fail to obtain funds at a rate you can afford.
- Assignment of responsibility to the conractor for obtaining building permits, inspection approvals, and complying with zoning and building laws. The National Association of the Remodeling Industry (NARI) recommends that you do not obtain building permits yourself. In most jurisdictions the person who obtains the permit is considered to be the contractor and is liable if the work does not comply with local codes. Let your contractor assume that liability, says NARI; it's part of what you are paying for. It's also the only way you can protect yourself.

  The contract should also include the price of permits, whether paid as part of the contract price or as a purchase allowance, sometimes used when the price of the permits remains to be established. If you're using a purchase allowance, stipulate that the contractor provide original receipts on a building department invoice form or letterhead.

  If you want to be present each time an inspector looks over the project, state it in this clause.
- Owner's right of supervision and inspection. This clause will establish that the work in progress must meet your approval and be subject to your inspection. It may appear to be in conflict with a standard, preprinted contract, which establishes the contractor's control over the job. Yes, the contractor is the expert, but if you have concerns about the way things are going, you don't want to see the job getting away from you. This amendment will prevent that; it also reinforces your right to express your concerns and opinions about the project.

- Completion dates. Include a clause that sets a date for substantial completion (meaning that the improved area is usable even though some minor work remains to be done before final completion) and establishes penalties for failure to complete on time, usually a deduction of a certain amount of money per day until substantial completion. You and the contractor will have to agree to a daily rate.

  Be aware that if you insist on a penalty for late completion, the contractor may insist on a bonus for finishing ahead of schedule.
- Project schedule. The date the contract is signed is the date of acceptance, which should be followed with all due speed by the date of commencement. Some delays are unavoidable; inclement weather and supply shortages, are two of the most common causes. Barring these, it is important that you ascertain that your contractor is available to commence work on the date listed in the contract.

  You and your contractor can work out the rest of the schedule. Some homeowners are content to let the job flow as it will and pinpoint only start and finish dates. (You *must* establish a finish date; an open-ended project, which could allow the contractor to take on other jobs while working on yours, might go on forever.) Any interim dates you set up will have to be flexible because renovation and construction jobs are fraught with surprises and unexpected delays.
- Daily starting and quitting times and hours during which workers have access to the site. The standard starting time for most construction is 8:00 A.M.; a project that starts much earlier than that will probably have your neighbors in an uproar. And to keep neighborhood relationships serene, let the people who live closest to you know that you are planing a renovation project, when it will begin, and how long it will last.

  Does the contractor want to speed the job along by doing some work on weekends? Don't agree to it until you have talked to your neighbors.
- Cleanup policy. Such a policy might read something like this: "At the end of every workday the crew shall remove

About Mechanic's Liens and Waivers

Perhaps the most grisly remodeling nightmare of all occurs when the contractor has not paid the subcontractors or has not paid for supplies, say, lumber, that have been ordered for your job and have therefore legally become part of your home. In these circumstances, you are liable by law to come up with the money.

If you fail to pay the amount due, which could be considerable, the subs and the supplier can sue you or slap a mechanic's lien on your home, preventing you from selling it until they have been paid. If you still fail to pay, you could lose your house. It's a stretch, but it could happen. "People don't realize that although they have dutifully paid their contractor they can lose their house if the contractor has not paid subs or suppliers," says attorney Ann Rankin, whose Oakland, California, firm specializes in construction law.

It's also a stretch that a contractor you have investigated so carefully would skip out and plunge you into this nightmare. But you must nonetheless protect yourself with a waiver of mechanic's liens clause, which requires the contractor to provide you with proof (or lien releases) at major stages of completion that subs and suppliers have been paid. Do not make final payment to the contractor until you have the waivers in hand. Rankin suggests increasing your protection by inserting a clause that requires the contractor to pay subs and suppliers promptly and indemnify the homeowner against liens.

In a large project with outside financing, the bank should verify that subcontractors and suppliers have been paid for each phase of the work before it releases funds for subsequent stages of the project.

In your state a waiver-of-lien clause may not be sufficient protection against the lien laws. Ask your lawyer if you need additional assurances, such as requiring that subs and suppliers also sign lien release forms stating that the contractor has paid them. And beware of lien sale contracts, which put your home up as collateral as a guarantee that you will pay the contractor even if his or her work is unsatisfactory in the end.

In some states suppliers and subcontractors may be required to send out Notice to Owner forms after they have been contacted to supply materials or labor for a project. Don't worry. These notices are not liens on the property, they are legal documents advising you of the suppliers' and subcontractors' involvement. Carefully compare the names of companies and individuals who send you Notice to Owner forms with the list of subcontractors and suppliers given to you by your contractor. If there are discrepancies, straighten them out right away.

dirt, dust, and debris as completely as possible, particularly if living spaces are involved. Tools and materials shall be stored out of the way as safely and neatly as possible." Rework this basic statement to suit your own needs. Your contract should also spell out that at the end of the project all debris is to be removed and the job site left broom-clean.

- Type, amount, and policy number of all insurances held by homeowner and contractor. As evidence of the contractor's coverage, insist on an insurance certificate from his or her insurer. This form will indicate that the insurer has been informed about the project and acknowledges its coverage.
- Dispute resolution. If the contract doesn't already contain a provision for dispute settlement, be sure to add one. A dispute-resolution clause assures that any dispute between you and the contractor that cannot be amicably settled will go first to mediation, then to binding arbitration with the service named in the contract. There are several ways to hook up with one of these services. Check your Yellow Pages, ask your lawyer, or contact the county bar association or the Better Business Bureau. (See Chapter 7 for details.)

  Some standard contracts contain language that limits the damages the contractor must pay to the cash price of the contract. Get rid of that kind of clause. Suppose you have contracted for a simple kitchen remodel at $15,000 and an open gas-supply pipe creates an explosion that not only demolishes the room but also puts you in the hospital. A $15,000 liability would hardly cover that situation.
- Cancellation, or breach-of-contract, clause. In the event of poor workmanship or inability or refusal to meet contract requirements, the homeowner must have the right to cancel.
- Waiver of mechanic's liens. Requires the contractor to provide you with proof (or lien releases) at major stages of completion that subcontractors and suppliers have been paid. (See box on page 106.)
- Warranties. This important provision should be carefully detailed, stating what it covers—labor, materials, or both—and for what length of time. (See box on page 109.)

- Details about any work to be done by someone other than the general contractor, including yourself. List such work here to make clear that it is to be handled separately from your contract with the general contractor.
- Change orders. Include a statement that the homeowner has the right to make changes to the project after the contract has been signed. Each change that is made should be accompanied by a mini-contract of its own with the details specified in writing and the document dated and signed by both parties. (See box on page 111.)
- Bathroom facilities. Will there be a lot of workers on-site? You may not want them all using your bathroom. In that case, specify that the contractor provide a portable bathroom for the duration of the job.
- Amenities. Add this clause if you want to clarify that you are not responsible for supplying water, ice, telephone service, or other amenities to the contractor, subcontractors, or any other members of the crew.
- Surplus materials. In this clause list any of the surplus materials or equipment that you might want to keep, and specify that the contractor remove all other surplus before the final payment is issued.
- Special instructions. Do you want the old appliances stored somewhere in your house instead of hauled away? Do you want supplies and materials kept in a certain place? Is any part of the house off-limits to the crew? Put it all in writing. If materials are going to be stored at the site, will they be covered by the contractor's insurance? Your homeowner's policy probably does not cover materials that have not yet been installed and become part of the house. Check with your insurance agent to determine your level of coverage on this point.
- The final payment. Make certain your contract specifies that you will not make final payment until the job is completed to your satisfaction in every detail, and to the satisfaction of the appropriate local authorities. You must thoroughly inspect the entire project and have all final releases of lien, in writing, in your possession before you write that final check.

About Warranties

Labor, the costliest aspect of any repair that will have to be made, is the coverage you should be most interested in. If a faulty cabinet shelf needs replacing, the lumber will cost you next to nothing, but the carpenter's labor could be high.

Try to get a full-year warranty of unlimited coverage for materials and labor from your contractor. Some experts suggest that homeowners negotiate for even longer warranties, especially on foundations and framing work. For roofs, five years is the minimum suggested warranty, and many roofs are warranted for 20 to 30 years.

Whatever the nature of your warranty, confirm when it goes into effect. Is it the date of completion, installation, or occupancy? In a limited warranty, also confirm what is omitted. In addition, you might also ask your contractor if he or she participates in extended warranties offered by a remodeling warranty company. For a fee, these companies guarantee the work for a significant length of time, provided they have satisfied themselves that the contractor has skill and financial solvency. This type of warranty is like an insurance policy issued by a private company, not by the contractor. They also offer a dispute-resolution service, and the warranty often stays with the house even if you move.

In most states implied warranties automatically go into effect when there is no written warranty. Although the details differ from state to state, the gist is the same: the contractor warrants that the work was done properly with quality materials. Implied warranties guarantee construction for anywhere from 1 to 10 years. Beware of clauses in your contract that might restrict the implied warranty offered by your state, which you can learn about through your state consumer protection agency. (See Chapter 7 for listings.)

If there are warranties on the products the contractor has obtained for you or on products you have purchased directly but have instructed the contractor to install, include them in this section of the contract.

Also, reserve as big a percentage of the total payment as possible for full completion of the work; the last payment must be significant enough to make the contractor stay around and finish the job satisfactorily. Twenty-five percent, according to some industry experts, is an ideal amount; 5 percent is too small, too easily shrugged off by a contractor who is ready to start the next job and needs a stronger incentive to finish yours on time.

- A provision stating that the general contractor cannot assign the contract to someone else.

Attorney Ann Rankin suggests a few more provisions: a clause that makes the contractor solely responsible for job-site safety, a prevailing-party attorney's fee clause (which means that if a dispute goes to court, the legal fees of the one who wins the judgment are paid by the one who loses), and an insertion that identifies the owner's representative or on-site supervisor, if there is one. (See Chapter 5 for more information on an owner's representative and what he or she can do for you.)

All these clauses and provisions probably seem overwhelming, and they may very well be more than your project requires. But be aware that more homeowner-contractor disputes develop in trusting relationships than in cautious ones; no matter how much you like your contractor now, you'll need protection for the disputes that may arise later. Go over these clauses with your contractor and your lawyer and choose only those that will assure smooth sailing and harmonious relationships. And *never* sign an incomplete contract or one that has blanks anywhere in it.

Keep the *original* contract and all attendant documents in your safe-deposit box or other inviolate place. Use copies to refer to as the job progresses.

A CONTRACTOR'S IDEA OF A GOOD CONTRACT

Some contractors are unscrupulous, we know; but so are some customers. One of the biggest fears that contractors have is that they won't be able to get the money fairly owed them from difficult or even larcenous homeowners. They generally want the money to come in slightly faster than it goes out, so that they are paid for work just before they have to pay for labor and materials. In the contract stage, the g.c. will endeavor to have the payment schedule go something like this—a small down payment on signing of the contract, next payment due when materials arrive at the site, other payments to coincide with the bills that he or she owes, final payment (which represents the contractor's profit) when the client signs off on the job. This schedule should not conflict with the homeowner's needs, but if it does, an attorney can help hammer it all out.

About Change Orders

Without protection, changes that you make in the project after the contract has been signed are potentially troublesome. Put all changes in writing—the details of the change, the agreed-upon cost, and the effect, if any, on the completion date. Sign and date this document and get the contractor's signature too. If the change saves you money, you want that fact on record so that you'll get credit for it. Well-documented change orders benefit the contractor too; if the change costs him or her money—say, substituting expensive paint for a cheap brand—the contractor will want to be compensated. No contractor is likely to switch to a more expensive product or material without protecting himself or herself, and neither should you. With change orders, always ask the contractor how much it will cost before you decide; a seemingly simple alteration to the plan, such as moving a door a few inches, could cost hundreds of dollars.

Although proper planning of your project can greatly reduce the number of changes you make, inevitably there will be some. You may change your mind about a paint color or the brand of a refrigerator; or you may decide that you want beaded board wainscoting in the bathroom instead of ceramic tile. Even if you don't change your mind about anything at all, the house itself may change it for you. The removal of walls or the pulling up of floors, particularly in an old house, may present you with surprises that require unexpected attention. You may find faulty wiring that must be updated, for example, or rotted floorboards that need replacing.

Contractors have several other concerns too, which is why they want to see some of the following clauses or provisions in the contract they sign with a homeowner.

- Prepermit work. Although contractors typically obtain permits for the project, they often want the contract to specify that the owner must obtain and pay for any easements, variances, or other allowances that may be required before a permit can be issued—and that if the owner fails to do so the contract is void.
- Site preparation. Many contractors will ask for a provision that releases them from the agreed-upon completion date if the homeowner has not prepared the site so that work can commence.

- Breach of contract. Contractors generally want a clause that includes the following: After 30 days of nonpayment, the contractor can submit notice to the homeowners that they have seven days to pay. If they don't pay by then, the contractor is free to pull his or her equipment and workers off the job and quit. In that case, the homeowner will be liable for payment of the work completed as well as for loss of time and materials incurred by the contractor.
- Finance charge. For late payments, contractors often charge 1 1/2 percent per month on the unpaid balance until it is paid in full.

Are boilerplate contracts okay?

Some contractors use preprinted forms with all of their customers, very possibly obtained from the American Institute of Architects (AIA), the Associated General Contractors (AGC), or the American Homeowners Foundation (AHF). The AIA contracts, the largest-selling standard contracts in the world according to an AIA representative, are widely used and recommended by professionals in the residential construction and remodeling industry. Homeowners also often use these documents. They are available in stationery stores and anywhere that preprinted contracts are sold. You can also get the forms directly from the AIA, AGC, or AHF. (See Chapter 7 for details.)

But not everyone is a fan of the AIA contract. N'ann Harp, for instance, thinks that it is not truly consumer-friendly and tends to "intimidate homeowners by its sheer interminable and indecipherable length." And, Ellis Levinson, a California reporter who covers consumer-protection issues advises, "Don't let a preprinted form intimidate you. Everything in these contracts is negotiable except for terms required by law." In other words, a standard contract may be an OK place to start, but to get the best protection, you will need to do some serious tinkering with it. "You cannot have too much detail in a contract," says Levinson. "Put in as many specifications as you can think of."

Although he thinks the AIA contract is "surprisingly evenhanded" Stephen Pollan suggests that it could use improvement in three critical areas—job commencement and substantial completion dates, payment schedules, and periodic inspections. If you decide to use the AIA form, you might want to ask a lawyer to beef up these areas.

If a standard form comes close to including what you need, you can use it and add the clauses that complete your protection. "We like the AIA and AGC forms because they include language that addresses the realities of remodeling or new construction," says Ann Rankin, "but we do modify them for individual jobs, adding, for example, a specific description of the project and other details. At this stage of the renovation process, it's very helpful to have an attorney to take care of these modifications."

You could also have a completely new, personalized contract drawn up by your attorney; but, says Rankin, "Why reinvent the wheel?" Some contractors will not accept owner contracts. If your contractor of choice abides by this philosophy, you may still be able to do business together by adding the clauses you are most adamant about to his or her preprinted version. Be sure, however, to go over this document carefully with your lawyer.

Smart Consumer Services issues a sample contract that can be used as is or modified to suit the homeowner and contractor. According to SCS, their model is a clear, easily understood document that demystifies legalese and educates the consumer while it protects. You can find a copy of that contract at the end of this chapter.

DO YOU REALLY NEED A LAWYER?

For a very small job, no. But for a major remodeling, yes. It's the smartest, safest way to proceed. "Having legal representation definitely prevents problems, and you don't need to spend thousands of dollars," says Ann Rankin, whose firm specializes in construction law.

In 1991, just after the Oakland, California, fire that destroyed thousands of homes, contractors, some honest and some not, flooded the area, looking for remodeling work. Rankin and her associates gave a seminar for the homeowners who were about to hire these people. "We covered the basics," says Rankin, "how to find a contractor, what to look for, how to check references, and the importance of a contract. Then for people who wanted us to represent them in their contractor negotiations, we offered a one-hour free consultation and a $500 flat fee." Some people refused the offer, mostly because they didn't want to spend the money; and, Rankin reports, a significant number of them came back later because they had gotten into trouble at the hands of unscrupulous or unskilled contractors, trouble that ended up costing them a lot more money than the original offer. Of the people who availed themselves of Rankin's offer, very few had problems, she says.

"What I see most often," says Rankin, "is a one-page-long, one-sided contractor's contract that is very weak; typically, it's got nothing about indemnification, payments, or work schedules. It's almost as bad as having nothing at all." Faced with something like this, you need an attorney to build in protections for you. One of the simple but invaluable things a lawyer can do is translate the items in a contractor's contract or a boilerplate preprinted one into everyday English that makes sense to you.

Many of us balk at the idea of hiring a lawyer. But one industry spokesman advises that if a remodeling job will cost you more money than you make in two weeks, you should protect yourself by having an attorney check any contract you sign. "Lots of people don't like lawyers to begin with," Rankin acknowledges, "and on top of that, they think that hiring one will be too expensive. They'll spend $400,000 or $500,000 to build a house but be reluctant to part with $1,000, or less for an attorney. But when it comes to a major remodeling, dismissing legal help is like failing to take care of your car. You can do regular maintenance and add oil when necessary, or you can rebuild your whole engine. Which is better?"

As long as you're spending the money for legal representation, you might as well get the best deal possible. Most experts think that would be a lawyer who is familiar with home remodeling and has written such contracts before. Rankin recommends finding a construction or real-estate lawyer or one who has actually worked in the field as an architect, engineer, or contractor. To find such a person, check with friends or relatives who have had work done, or ask your local bar association for the names of a few lawyers who specialize in construction or home remodeling.

If you're determined to work without a lawyer—which in this situation is somewhat like an inexperienced trapeze performer working without a net—you can still get some helpful advice about the contract that you and your g.c. have devised. One source of help is the contract advocacy review offered by Smart Consumer Services. This low-cost ($49.95) alternative to a full-scale legal review examines *un-signed* contracts that a builder or remodeler has presented to a homeowner, highlighting such items as misleading language, excessive payment or deposit requests, and other red flags. The contract review also points out important questions the homeowner ought to raise with the contractor and recommends ways the owner can request changes.

"A contract review by an attorney or by the Smart Consumer Services contract advocacy service—which, by the way, a lender will probably pay for—is simply a no-brainer way to insure that everybody knows exactly what to expect, when, and for how much," says N'ann Harp, the guiding light of Smart Consumer Services. "These days it's just plain asking for trouble not to have home-improvement contracts reviewed. The statistics of complaints prove it.

"Our review looks for several things," Harp explains. "Is the contract clear? Is it complete? Is it fair? We also make homeowners aware of missing, ambiguous, or conflicting information and remind them of their rights and options when negotiating. In addition, we advocate clarity in the construction documents. Our advocacy review can help reduce the chances that miscommunication will be a source of conflict between a contractor and homeowner." (For details, see Chapter 7.)

According to RemodelingCorner.org, a new web site geared to help ing homeowners establish a positive relationship with contractors, there are three good reasons to protect yourself with a contract review:

1. If something goes wrong, the odds are against you. Statistically, the homeowner is most likely to lose if some mid-job dispute or problem arises and the contractor leaves with the customer's money. You may like and trust your contractor, but don't be in denial about what can happen.
2. A good contract allows you to retain control. Customers need a precise schedule of completion milestones and payments so that the contractor does not have excessive amounts of their money at any given time. A good contract spells out these things, and many others, in great detail so that everyone knows from the beginning where he or she stands and what's expected.
3. A contract review is cheap insurance. Think for a moment about how much money you are spending on this job and what percentage of your annual salary it rep-

resents. Compared to those figures, a contract review is a bargain. It will clarify any overlooked questions, spot red flags, and give you additional reassurance that you've made a sound hiring decision.

BUILDING PERMITS

First and foremost, do not apply for any permits yourself. Your contractor is legally responsible for the proper completion of the work and should obtain the permits in his or her own name. Be wary of any contractor who asks you to get the permits.

Even though you will not be obtaining these documents yourself, you will have done your homework as outlined in Chapter 2, and you will know exactly what sorts of permits you will need.

Once a permit is issued, your project is automatically registered with the local building authority, which will inspect the work while it is in progress and after it is completed to establish that it has been done to code. If any of these inspections reveal that the plan has been altered in a way that conflicts with the building code, your project will be suspended until you correct the flaw.

APPLICATION AND ISSUANCE

Your contractor will need the final blueprints before applying for permits. The local building authority studies these plans to see what the project entails, so that the appropriate permit can be issued and inspections can be conducted.

Permits are issued on first-come, first-served basis and can take anywhere from two to six weeks to be processed. No, there's nothing you can do to make it go faster, although extraordinary circumstances occasionally speed things up. But your project can proceed while you wait, provided no actual construction is done. For example, in a kitchen remodel, your contractor could remove old cabinets and prep the wall behind them, but would not be able to install new cabinets. Be careful about this little rule. Breaking it could cost you as much as $500; and if your town officials are really tough, they can make you remove all new work and begin again when the permit comes through.

Get Wise About Warranties

Will you be buying some major appliances as part of your remodeling project? If so, check the warranties before you make a final decision. If your contractor is purchasing major items for you, check warranty coverage with him or her before you agree to the purchase. **Full warranties** offer the best protection and require the warrantor to fix a problem without charge throughout the period of warranty. For appliances, warranties often run for one year and cover parts and labor. **Limited warranties** can stipulate almost anything, as long as the provisions are stated in a document that is available for review at the time of purchase. **Multiple warranties**, which combine aspects of both, may offer full coverage for specified parts of a product while offering limited coverage for other parts of the same item.

Most appliance dealers will try to sell you extended warranties. Read the fine print carefully. Some policies extend indefinitely, at an escalating cost, while others end after five to seven years, just when problems are more likely to occur. Also, weigh the long-term cost of the warranty against the likelihood and cost of a repair. A warranty covers defects, which generally appear the first year, not normal wear and tear.

If you have a complaint that has not been resolved to your satisfaction, you might get some help from the Major Appliance Consumer Action Panel (MACAP), an industry organization that mediates customer complaints that have not been resolved by either the dealer or the manufacturer. (See Chapter 7 for details.)

Meanwhile, remember these tips from MACAP for maximizing warranties:

- Test all the features and controls as soon as possible after purchase. If you discover a problem after the warranty has expired, it is unlikely that you'll be covered.
- Keep detailed receipts for purchases and service. If your 11-month-old refrigerator is repaired under warranty, and the same component needs repair two months later, the manufacturer may handle the repairs if you can prove that it is a persistent problem.
- Put all complaints in writing and describe in detail the problem, date of occurrence, people you talked to (get their names!), and the actions that were taken.
- To prevent voiding a warranty, work only with the authorized service dealers listed in the manufacturer's information booklet. (Beware: Just because a service center advertises itself as repairing certain specific brands does not mean that it is an authorized center.)

The red tape involved in procuring a permit can be annoying and frustrating, but don't delude yourself that you can slide out of getting one. Code enforcement officers are usually very vigilant and will probably catch you in your little subterfuge. If they do, you may have to redo some of the work, say, remove the wallboard so that inspectors can check the wiring or plumbing that has been added; in a worst-case scenario, you may even have to demolish the construction entirely. More bad news—without a valid certificate of occupancy, which the building authority issues, you cannot sell your house.

Do remember that your town or county may require a permit even if you are doing the work yourself. Check it out before you begin.

PERMIT FEES

When calculating your projected remodeling costs, it's easy to forget some of the expenses that occur along the way. Permit fees, for example, often don't make it into the final tally, though they can sometimes reach into the thousands of dollars. Check with your local building authority to learn precisely which permits your project will require.

Although permit requirements vary widely from one region to another, some of the most common include the following:

- General building permit, often based on the cost or the scope of the renovation or addition.
- Plan-check fee, for the time the local building authority spends checking your plans for compliance with community codes.
- Tap fees, to cover the cost of tapping into municipal water and sewer systems.
- Driveway or street-cut permits, to allow you to install a driveway or cut into an existing curb to relocate a driveway.
- Impact fees, which cover the effect your renovation has on roads, law enforcement, fire protection, schools, and other community services.

After You've Signed: The Three-day Cancellation Rule

Federal law states that you may cancel any contract valued at $25 or more provided you do it within three business days. You may also cancel contracts that will be paid in installments for more than 90 days. Same three-business-day rule applies here.

In many states the law requires contractors to inform you of these rights and to furnish you with a notice-of-cancellation form. Should you decide to cancel within the three-day period, send your cancellation notice to the contractor by certified mail, return receipt requested. And keep a copy.

If you wake up in the middle of the night in a cold sweat regretting for some reason or other your decision to hire a this particular contractor, convinced that you made a really poor choice, take advantage of this loophole. It's better to be a little embarrassed than to live with a bad decision.

SAMPLE CONTRACT

www.SmartConsumerServices.org

1. Parties:

Homeowner________________________________, whose home address is (City/ST/ZIP)_ ________________________________, desires to hire Contractor ____, whose company, ________________________________,is located at (City/ST/ZIP) ____, to perform herein described work at the following address and location: ____.

NOTE: Be sure to include the Lot # or legal description of site where the work will commence, especially if that address is different from the owner residence or if the property is a second home or investment property.

2. Job Description

____

____

NOTE: This should be one paragraph describing the scope of the project. It should reference any related documents, such as dated blueprints, drawings or plans provided by a named architect or firm, which will define the work more accurately.

EXAMPLE: "Project will include demolition & remodel of existing kitchen; powder room installation beneath entry stairway; construction of an approximately 700-sq.ft. addition/family room to the rear of residence; and roof/siding repair, according to specifications provided by XYZ Architects' plans dated 3-1-98".

3. Permits & Approvals

Approvals: Homeowner accepts responsibility for obtaining any necessary approval for work to be performed from local homeowners association (HOA).

NOTE: Written permission may be required from the Board of Directors of the HOA for any exterior work, including roofing, paint colors, siding, fencing, paving, and landscaping. Do not assume that any work will be "too minor" to draw the attention of the HOA Board. Allow months of advance notice for approval, with drawings and planned work submissions, including color samples. When in doubt, re-read the CC&R (Covenants, Codes and Regulations) documents received at purchase.

Permits: Contractor accepts responsibility for determining the need for and obtaining any building permits that may be required to perform the specified work.

NOTE: ONLY contractors should handle acquisition of any and all building permits. Under no circumstances should Owners allow themselves to be convinced or cajoled into "doing the Contractor a favor" or agree for any reason, to personally sign for any required building permits. Why? In the event work done on Owner's premises fails to pass building inspection the only party held legally (Read: financially) responsible to bring the building " to code" will be the one whose name is on the permit. Abusive contractors can do below-code work and walk-away without any responsibility...if an Owner signed for the building permit

4. Licenses, insurance & bonding: (Not required in all areas.)

Contractor agrees to comply with all state and local licensing, insurance, worker's compensation and registration requirements.

Contractor's state license and/or local registration is for:
what state____________________License #______________________________
what jurisdiction________________License#______________________________
what type of work____
Contractor has no license for the following reasons:
____

NOTE: Find out in advance from your state & county if licensing is required for all or what sorts of home improvement firms, builders, landscapers or handyman companies.

Workers' Compensation Insurance copy has been provided

Policy Number____
Agent Name ____
Phone____
__ Bond: Jurisdiction ______________________________ Amount $__________
__ Copy of current bond policy has been provided
__ Liability policy in the amount of $__________________.

NOTE: Bonding is required in some, but not all, areas. Contact your county, city or township government offices to determine if a contractor, tradesperson or handyman should provide proof of adequate bonding and in what amount. In areas where bonding may not be required it is still essential that a contractor carry liability insurance to protect against possible loss or injury resulting from an on-site accident.

5. Materials List

All materials shall be new, unless otherwise specified, and of good quality, in compliance with all applicable codes and laws, and shall in all appropriate cases be covered by manufacturers' warranties, which will be provided to owner.

___ Materials shall consist of: ____________(Materials list attached separately)

___ Homeowner shall purchase the following materials: _________________

____

___ Contractor will order all materials but owner will pays suppliers directly upon delivery.

NOTE: Direct payment of suppliers protects Owners from a possible supplier's lien in the event the contractor fails to pay for materials used on the job. Direct payment of suppliers is also a method of confirming materials costs in a Time & Materials Agreement.

Materials List should include grades of supplies or raw materials, brand names, model numbers, colors, styles, etc. of all products, appliances, equipment installed or used.

6. Warranties

Product Warranties have been provided for the following:

____

____

Limited Warranties: Contractor agrees to complete the job in a workmanlike manner in compliance with all applicable building codes and according to standard practices of the trade.

Labor & materials provided will be free of defects for _______ yr(s) from completion, or correction, repairs or replacements of any such defects will be provided at no additional charge within that period.

Additional warranties provided by the contractor:

____

NOTE: Some states require contractors to provide minimum one or two-year limited warranty periods on workmanship. Find out in advance what, if any, minimum warranty period you are entitled to receive. Unscrupulous contractors have been known to include shorter warranty periods than that provided by their state's law. Uninformed homeowners can actually sign-away their own consumer rights.

7. Liens & Waivers

Contractor agrees to protect Homeowner from liens being filed by any contractor, subcontractor or supplier of materials for this job in the following way:

___ Contractor will provide Homeowner with acknowledgment of payment in full or waiver of liens, or release of liens from each subcontractor or materials supplier prior to final payment for the specified work.

___ Contractor shall provide Homeowner with lien waiver or release copies from any subcontractor prior to hiring him/her for the specified work.

___ Contractor shall not use any materials without presenting Homeowner with an acknowledgment of payment in full by supplier.

___ Contractor and Homeowner agree that Homeowner will be protected from supplier or mechanics' liens in the following way:

____

NOTE: Lien Release and Waiver of Lien forms are provided with the Smart Consumer Services Sample Contract Package, as well as an excerpt from The Smart Consumer Guide to Working with Home Improvement Contractors on the topic of liens and why it is important to include specific language in the Agreement, and to address the issue openly with a contractor prior to hiring.

8. Waiver of Liability

Contractor agrees to exempt Homeowners from liability for any injuries sustained by Contractor during performance of specified work.

9. Dispute Resolution

___ In the event that a dispute occurs between Homeowner and Contractor that cannot be amicably resolved, parties agree to select a third party mediator who is mutually acceptable and neutral to the situation, to help resolve the problem, for whose services they will share costs equally.

In the event mediation attempts are unsuccessful the parties agree to:

___ Submit the case to an arbitrator subject to the rules of the American Arbitration Association, whose decision will be final

___ Submit the case to Small Claims Court if the amount is within the jurisdiction of the court

___ Settle the case according to applicable state laws

NOTE: Smart Consumer Services recommends the use of mediators specifically trained in Alternate Dispute Resolution. A local chapter of the American Bar Association can help you locate an ADR-trained mediator.

10. Time of Performance

The work will begin on ________________ (date, Year) and shall be completed within an anticipated _____ (hours, days, weeks) but no later than ________________(date, year). Time is of the essence.

NOTE: In commercial construction contracts it is not uncommon for there to be penalty fees charged to Contractor for each day a project is delayed beyond the indicated completion date. If Owner feels strongly enough about the need to be completed by a particular date it is not unreasonable to negotiate either a penalty for lateness or an incentive for early completion of a project.

11. Time & Materials Options

The described work will be performed for a total labor amount of $__________,
or be done at the rate of $______per hour. Total labor not to exceed
$__________.

___ Homeowner shall pay for materials upon delivery to the job site

___ Homeowner shall reimburse contractor for materials' receipts

___ Labor payments shall be made on a ___________ (daily, weekly) basis.

Final labor payment of not less than $__________ shall be withheld until satisfactory completion of the job.

NOTE: Smart Consumer Services recommends that Time & Materials Agreements are entered into for small repair or handyman jobs. There is too much incentive for an unscrupulous contractor to waste time when the-longer-it-takes-the-more-they-make. Most jobs are best put out for bid.

12. Deposit & Milestones of Completion Payments

Contract total $___________________________ for all labor and materials.
A deposit equal to ________% of contract total. Deposit $_____________
__ No deposit required
Amount due upon completion of ___________________: $_______________.
Amount due upon completion of ___________________: $_______________.
Amount due upon completion of ___________________: $_______________.
Amount due upon completion of ___________________: $_______________.
Balance on satisfactory completion, including any Punch List items:
$____________.

DEPOSIT: Most remodeling jobs should require a deposit amount of no more than a 10%-15% of the contract total, unless there are special order materials or custom cabinetry, for instance, that cannot be returned or will take time to produce as a one-of-a-kind feature. A request for a large deposit should be looked at as a proceed with caution indicator. A company that asks for no deposit is usually one that has good cash reserves, has good credit with suppliers, and pays its bills on time.

MILESTONE or PROGRESS PAYMENTS: On a lengthy or complex job all payments should be tied to specific "milestones of completion" that indicate the work is progressing properly. It is smarter to make smaller, more frequent payments than to agree to a few, large ones. In that way Owner retains more control over the funds and gives an unscrupulous contractor less opportunity for abuse...or worse, to abandon a job immediately after a large payment.

FINAL PAYMENT: One of the reasons so many homeowners can't get contractors to finish punch list items after the final payment is because the contractor has no further financial incentive to return. Owners have given away their last bit of control. Insist that final payment be made only after the Punch List items are 100% completed. Also insist that Final Payment be an amount between 15%-20% of the total contract amount, which is the normal percentage of profit a contractor has built into the contract. Work out the Progress Payment schedule backwards from a modest Deposit and a substantial Final Payment.

13. Change Orders & Amendments

Homeowner and Contractor agree that any modifications or changes of this contract in cost or scheduled work described herein shall be in writing, co-signed by Owner & Contractor and shall be amended to this contract as dated Change Orders.

___ Homeowner and Contractor additionally agree in (a) separate attachments labeled

Attachment A, that:____

Attachment B, that:____

Attachment C, that:____

14. Cooling Off period

homeowner and Contractor understand and agree that, by U.S. Federal Law, the owner has three days to withdraw from this signed agreement, without penalty. No work will commence on this project until the three day period after signing has expired.

Signed

Homeowner ____________________________________ Date_____________

Contractor ____________________________________ Date_____________

Provided by Smart Consumer Services,
a consumer education and assistance organization.
www.SmartConsumerServices.org

Release of Lien

Known All Men by These Presents:

the undersigned,____

for and in consideration of the sum of $ ___________________, and other good and valuable considerations to the undersigned in hand paid , the receipt whereof is acknowledged, does hereby waive, release and relinquish the undersigned's right to claim, demand or impose a lien or liens for work done or materials furnished or any other kind of or class of lien whatsoever on the following described real property in the county of
________________________________, State of _______________________,

Description of real property:

____

____

Dated this _______________ day of ____________________ 20______

Signature ____

Address ____

____

Provided by Smart Consumer Services,
a consumer education and assistance organization.
www.SmartConsumerServices.org

Waiver of Mechanic's Lien

Known All Men by These Presents:

Whereas, ____, the undersigned,of:

____company, city of,____. County of: ______________________________ State of: ________________________, has been employed by: __________________________________ under an agreement to furnish ____ for the building known as:

____.

Now, therefore, let it be known that______________________________, the undersigned, for and in consideration of $ _________________, and other good and valuable considerations, the receipt of which is hereby acknowledged, does hereby waive and release any and all lien, or claim or right of lien on the above described building and premises on account of labor or materials or both, furnished or which may be furnished by the undersigned to or on account of the agreement for the building or premises.

Dated this _______________ day of _____________ 20_______

Signature ____

Address ____

Provided by Smart Consumer Services,
a consumer education and assistance organization.
www.SmartConsumerServices.org

CHANGE ORDER

Customer Name: ______________________________

Job Location: ______________________________

Requested Modification: ______________________________

____

____

____

Attachments: ______________________________

Cost of Modification $______________________________

Estimated change to completion date ______________________________

Parties agree upon co-signing that this Change Order has become part of the originating document and contract for work in-progress.

Approved by:

Owner: ______________________________

Contractor: ______________________________

Date: ____________________

Provided by Smart Consumer Services,
a consumer education and assistance organization.
www.SmartConsumerServices.org

five

The Job Begins

The contractor has been hired, the contract signed, and the date set for work to begin. You've put a lot of time and thought into your home-improvement project already. Can there really be more work to do? Well, yes; but not much.

You're almost ready to begin, but you do need to take care of a few important details before the big day dawns. And to make sure that you and your contractor remember what those details are, get a nice big notebook and record everything that is discussed and decided from now until the end of the project. Some things, most notably change orders, must be treated more formally, as spelled out in Chapter 4; but for many details of the job, a notebook is adequate. If there is ever any question about what was decided, you can leaf through the pages and find a record of the discussion or the item in question. For example, after the countertops and backsplash are installed in your kitchen, you realize that Cameo white paint is going to look much better on the walls than the Navajo white you had specified. Tell your contractor about the change when you see him that day or, even better, hand him a note. You probably don't need a change order for this, but you should definitely record the date and the details of the change in your notebook and have the contractor initial it.

Another important detail: be certain that the contractor has obtained the necessary permits. Construction proper cannot commence without the permits, although the crew might be able to do a little preparatory demolition work while you wait. But you don't want to wait, so take the time now to clear this up. Remember, it's the contractor, not you, who obtains the permits.

WHAT'S IN STORE? FIND OUT!

You have a clear idea of what your project entails. You also know by now that construction and remodeling are not exact sciences or high-tech processes. On any job there are problems that could not have been predicted but can get solved by a skilled and experienced contractor. There are also delays and mistakes. Get ready to go with the flow.

If you're not clear about what the project entails, discuss it with your contractor now. If your renovation involves an addition to your house, the first steps will probably be site preparation, excavation, and foundation work. Then come rough carpentry to frame the addition and a roof to cap it off. If you're remodeling existing space, demolition—a loud, messy, but usually short process—will come first, followed by rough carpentry, insulation, installation of windows and doors, and wallboard. Plumbing and wiring may also be involved. Then comes finishing work, which could include cabinets or built-ins, new floors, painting, tiling, trim work, phone lines.

Once you realize what's in store, take a realistic look at how stressful, almost traumatic, this process can be and how much it will disrupt your household. No matter how many precautions you take, your house will be somewhat, or even very, dusty and grimy for the duration of the renovation and for weeks thereafter. Rooms where you've felt cozy and comfortable will no longer be accessible, and even the rooms you are able to use may be crammed with furniture and other objects you've had to move from somewhere else. Your furniture will be shrouded in old bedsheets and plastic sheets. Your front and back yards will be filled with unsightly equipment and stacks of lumber, your driveway clogged with strange vehicles. You will feel invaded and you will miss the comfort and familiarity that you have come to expect in your home. Strangers, like an occupying army, will be walking through your rooms talking, laughing—at times, even shouting—and looking quite at home. There will be lots of noise—the crashing of demolition debris, whining of saws, banging of hammers.

The best thing you can do is get out of the way of this noisy crew. It may look like chaos, but they know what they're doing, and they need space in which to do it speedily and efficiently. They will not be neat while they're working, although by the end of the day they will have cleaned up the site to your satisfaction.You specified this in your contract, remember? They last thing they want is you tiptoeing in and out, wringing your hands, stepping over boards, getting in the way, and posing a safety hazard.

If your whole house is going to be seriously torn up, you might want to move out for a while. Ask your contractor how disruptive he or she thinks the project will be for day-to-day living; if it sounds like the conditions will be especially trying, think about staying with friends or relatives, renting a house for a short time, or even staying in a motel. Work will go appreciably faster in an unoccupied house, and you'll be sparing yourself a great deal of stress. However, don't go so far away that you're not able to monitor the job daily.

You might also want to set up a mini-office. You won't necessarily need a desk and a chair, but you should establish a safe, centrally located place for your project-related files. You will be receiving a steady stream of documents and bits of information for the duration of the job. For smooth sailing, and for your own sanity, keep the following documents together in an organized and efficient file:

- The contract
- Payment and project schedules if not part of the contract
- Construction drawings and floor plans
- Product and materials specs
- Correspondence between you and your contractor
- Other correspondence or agreements with third-party participants
- Paint chips, manufacturer's samples, product literature
- Warranties
- Change orders

- Lien releases
- A list of questions to ask your contractor
- Notes and reminders to yourself, including daily decisions you and the contractor make

Before work starts, gather your family and let them know what the process entails and how long it is likely to last, suggests the National Association of the Remodeling Industry (NARI). For a certain period of time, your home, your day-to-day routine, and the conveniences you are accustomed to will be disrupted. A stressful situation such as this will be much more manageable when your family knows what to expect and can prepare themselves for it. NARI also advises that you set up some strategies for dealing with the mess and stress. Discuss what each one's role and responsibilities will be during the project so that you can approach it as a family partnership. Looking at the remodeling as an opportunity, rather than an imposition, can bring everyone together with a common goal.

For instance, you might ask the children to be responsible for securing the pets and keeping other kids away from the construction site. This will give them a sense of involvement. And let everyone in the family have a say in the many decisions that will have to be made during the course of the job. Everyone will have to deal with the dirt and the noise, so you might as well let them also take part in the decision-making process and the final result, says NARI. And keep reminding everyone that this is creativity in progress. It's dirty and disruptive, yes; but the results will be worth it.

To stay as serene as possible in the midst of potential chaos, try getting away now and then for a long walk or a leisurely dinner out. In the evening, when the workers are gone, play soothing music or luxuriate in a hot bath, or both. If your bathtub has been rendered unusable by the makeover, see if one of your friends will let you use their tub for an hour or so.

PREPARING FOR THE INVASION

No matter what a good job you have done with preparations, you probably did not think of everything. Your contractor probably didn't either. Review some of the following housekeeping details with him or her before work starts:

Contact person. It's probably the contractor, unless you have another on-site supervisor. Do you have all the telephone and pager numbers you need to contact this person in an emergency? Whom do you call if he or she is unavailable?

Amenities. If your contract restricts the use of telephone, kitchen, and bathroom facilities for contractor and crew, remind the contractor of that provision. If you didn't specify, make some rules now and record them in your notebook. Make clear which bathroom the crew can use and restrict traffic to that one only. If you don't want the workers trooping in and out of the house to use your bathroom, the contractor must provide a portable toilet and you must decide where it should be placed.

Construction workers often play radios—loudly, sometimes—while they're working. Is that OK with you? If not, talk to the contractor about it.

Access. If possible, designate one door to your house as the contractor's entrance and give him or her a key for access when you are not at home. If you have an alarm system, you can assign the contractor a temporary code, then change it when the work is done; this will keep the house protected by an alarm at all times.

Also, discuss security measures for times when you are not at home. Who will be responsible for locking doors and windows at the end of the work day?

Parking. Is your driveway large enough to accommodate construction vehicles and your own car or cars? Will you need to get your car in and out during the day? If so, make it clear to the contractor that you don't want work vehicles blocking your way.

Messages. Establish a specific place to leave messages for the contractor. It's a good idea to choose an out-of-the-way place so that only the contractor has easy access to it, which will be helpful if you are having a problem with a subcontractor or worker or just want to communicate something private to the contractor.

Dust management. If existing rooms of your house are being renovated, there will be dust in those rooms *and* in the spaces that are not being worked on, and it will creep up and down stairs too. If your project is an addition to the house, the dust problem will not be so bad. Discuss with your contractor how to best protect your furniture and other household objects. If the method he or she employs doesn't seem adequate—remember, this is very fine, insidious dust that gets into everything—come up with some ideas of your own. For example, drape bedsheets or plastic sheets over furniture; put plastic over doorways too—professionals recommend that a 4-mil thickness of plastic be taped snugly over the doors to all rooms that are not being worked on. You can also roll towels into cylinders and lay them along the bottom of each door to keep dust from sneaking underneath.

To protect the floors of rooms that workers will pass through, cover them with drop cloths. Tape plastic over air-conditioning or heating ducts in any room that will get dusty.

Boundaries. If you are going to live in the house while the work is being done (see more about this below), you will want to designate some areas for family use only. Discuss this with the contractor and request that the crew be told which spaces are off-limits.

Work zones. The crew will need an area (maybe several) where they can saw lumber, mix paint, and store tools and materials. It might be the garage, the room or rooms being worked on, or in good weather, even the backyard or patio. You and the contractor should make this decision before construction begins.

Safety issues. You'll need to declare the construction zone off-limits for children and pets. It sounds obvious, but many homeowners overlook it. Don't make that mistake; construction sites are dangerous places. Caution your children about this and keep your pets somewhere safe until the project is finished.

First things first. If you want certain parts of the job completed before others, ask the contractor to arrange it if at all possible, and be prepared for the fact that it might not be possible. You may want the nursery finished first so that you can be decorating it for the new baby while the crew works elsewhere. Part of your project may be a new, or remodeled, home office, which you will probably want to have done quickly so that you can get back to work.

Even if you have no special requests, find out where the crew will begin the job so that you can get those areas prepared for their arrival.

Major disruptions. To minimize the disruptions that the project is bound to have on your household, ask the contractor to let you know ahead of time when major demolition will take place or when electricity or plumbing will be out of commission. You'll also want to know when applications of polyurethane or other foul-smelling solutions will make certain rooms—or the entire house—uninhabitable. This might be a good time to stay with friends or at a motel for a night or two.

Holidays. Your contract specifies workdays and start and finish times, but what about holidays that occur during the project? If the contractor wants to work on those days, will you agree?

WHAT ABOUT THE FURNITURE?

If existing rooms of your house are being remodeled, you will have to get the furniture out of those spaces and into safe storage of some kind. In some cases it may be sufficient to move your furniture to the center of the room and cover it with sheets or tarps. It's a good idea to do this before contractor and crew arrive. You don't want to be paying them for moving furniture, and they certainly don't want to be doing it. If there are one or two very heavy items you can't mange yourself, you can ask the crew to move them; but do as much as you can yourself. If it's too much for you, enlist the help of friends and neighbors, or hire local movers for a couple of hours' work. Special pieces, such as a grand piano or a treasured antique, could require the services of professional movers and maybe storage as well. Storage lockers hold a lot for a fairly reasonable rate.

Give some thought also to where you furniture will end up when you move it out of the way. Perhaps you were planning to stow it in the garage only to find that the work crew needs that space. Better discuss this issue with the contractor before Day One of the remodeling.

Get small things out of the way, too—paintings, rugs, lamps, dishes, knickknacks, books, magazines, toys. It's annoying to have to pack them up and store them away, but doing so will guarantee the safety of objects that you treasure and will allow the construction crew to get right to work. Remove breakables from walls and shelves in and around the construction site, not only in the room that's being worked on but in adjoining rooms too and maybe even upstairs. Consistent heavy banging with hammers or demolition tools can knock treasures right off shelves. When you're getting a room remodeling-ready, don't forget to leave out whatever you will be needing for the duration, such as the TV set, telephone, houseplants, coffeemaker, dishes, pots and pans, towels, toiletries, and the book you're currently reading.

Kitchen renovations pose some genuine challenges—you need to move a lot of items out *and* find a place for a temporary kitchen. You could eat out every night, of course, but that might not be a sensible option for the six to eight weeks that the renovation could take. A combination of some meals eaten out, some brought in from take-out restaurants, and some prepared at home is what most people choose. Keep the temporary kitchen simple. A coffeemaker and toaster could take care of breakfast; a hot plate, electric cooking pot, and microwave can produce a perfectly acceptable lunch or dinner. If you have access to the refrigerator, great (the crew might be willing to move it into your makeshift kitchen). But if you don't, a cooler stocked with a few bags of ice makes a nifty emergency refrigerator. Place this little setup as far away as possible from the job site, even using the corner of an upstairs bedroom if necessary, and as close as possible to running water. Proximity to an upstairs bath, downstairs powder room, or basement laundry sink will make everything easier.

PROPERTY RITES

Your property may undergo stress and trauma too. If your remodeling project involves building an addition, some major excavating equipment will soon be lumbering into your yard, ready to tear everything up. How will you protect your shrubs and flowers from it and from the ladders, workers, and debris produced by the installation of siding or a new roof? What about trees? Will any of them need to be moved to escape the claws of the excavating equipment? And any sizable renovation, whether it involves an addition or not, will very likely require a Dumpster to hold construction debris. Where will the Dumpster be put so as to cause the least amount of disruption in your yard?

If you haven't already considered all of these things and written instructions for the handling of them into your contract, do so now. To keep shrubbery and trees safe, contact a landscaper, who may have some suggestions about protecting or temporarily removing them. This will cost you a little money, but will still be appreciably less expensive than replacing your old, treasured trees and shrubs. Flowers in beds near the spots where work will be done can be easily saved.

Just dig them up and give them a temporary home in another part of the property.

Your first step is the Dumpster challenge. Find out if it can be placed at the curb in front of your house. If local regulations disallow that, what about the driveway? Is it large enough to hold the Dumpster, construction vehicles, and your own car? If your driveway is too small, can construction vehicles park on the street or can you use a neighbor's driveway during the day? The sight of a Dumpster sitting there might inspire you to use it for the household purging that often accompanies a remodeling; but first, find out if you can dump non-remodeling-related trash in it.

Some communities or neighborhoods prohibit Dumpsters, in which case debris is piled in the yard and covered with a tarp until the contractor hauls it away.

DAY ONE

The day you have been planning for finally arrives, and you face it with an odd mixture of excitement and dread. Both feelings are appropriate at a time like this.

You've met the contractor several times, but the rest of the crew, which will probably arrive on your doorstep at eight in the morning, will be new to you. They'll look strange and slightly scruffy, and it may dawn on you that you will be sharing your house with these unkempt strangers for a while, a few weeks at least, maybe months. It's an unsettling thought. What do you do? Greet them? Ignore them? Hide from them? Offer them breakfast?

Surprisingly, there is some disagreement among building professionals when it comes to the attitude customers should adopt toward contractor and crew. One contractor of our acquaintance says, "The best way to establish control over the project is to maintain a business, not a personal, relationship with the contractor. You will want to have a good working relationship, yes; but it is not your job to provide cookies and milk. They'll just think they can take advantage of you and, besides, as soon as a payment is late they'll forget all about your generosity." Other professionals advise taking a much friendlier stance. Sven Swenson, for example, believes that only by taking a friendly and cooperative attitude toward the contractor can a homeowner successfully weather the storms of the remodeling process. "Welcome the g.c. and crew to your home. Greet them pleasantly each morning. Offer them a cup of hot coffee or a glass of ice water from time to time. You don't have to give them your firstborn son; just be friendly. If you create this kind of relationship from Day One, you will be much better able to work things out should there be a problem or even a crisis," Swenson predicts.

Keep Remodeling Money Separate

Because you want to be able to figure out at a glance where your remodeling money is going, you might decide set up a separate bank account strictly for the project. This is especially important if the renovation is a large one that will take more than a couple of weeks to complete. Savings or checking, it doesn't matter what type of account. What does matter is that your remodeling money is not intermingled with other funds and that you have a clear picture of what you are spending. However, if you choose a savings account, use money orders or bank checks for your disbursements so that you have proof of payment.

If you are financing the project with a home-equity or home-improvement loan, the money will probably be issued to you in a special checking account, which is ideal for tracking your spending. Otherwise, just open a new account and designate it for the project only.

But you must maintain enough of a businesslike attitude to be assertive and effective when necessary. You have done this to some extent in the interviewing and bidding processes and in contract negotiations. Now that you've chosen a skilled, responsible person who's going to do a good job for you, you may think your troubles are over. Probably not. Mistakes happen, problems arise, and conflicts emerge. At these times you'll need to be assertive and able to stand up for yourself.

Repeat this little litany, or something like it, to yourself when you feel nervous about confronting the contractor or when the contractor responds to your questions or requests in a dismissive, uncooperative, or angry way: "This is my house. This is my project. I am paying this person with my money. I am treating this person with fairness and respect. I am entitled to get what I want unless it is for some reason impossible to achieve. I am certainly entitled to get what I agreed to and have paid for. I am also entitled to be heard." You're not saying this out loud, mind you, just using it as self-talk to cheer yourself on.

DELIVERIES

As soon as the job gets under way, materials and supplies of many kinds will start showing up at the site.

If you're having a foundation dug and poured, this parade of products will begin with excavating equipment; then perhaps concrete blocks or premixed concrete; or cement, sand, and gravel to the mixed on-site. If there's to be no digging, the lumber will start arriving, maybe stacks and stacks of it.

Do not sign for any of these materials. That is the contractor's job. Materials are delivered to him or her, not to you. If you sign, you are agreeing to accept and be financially accountable for supplies you may know nothing about, supplies that may not even be for you. If the contractor is not on-site when supplies arrive, try reaching him or her on the phone; but do not step in and try to make his or her life easier by signing anything. By the way, the contractor *should* be on-site when deliveries are made. If he or she is not, find out why and make clear that you will not be accepting any deliveries for the job.

It is permissible, however, to get nosy about the materials. Ask what the delivery consists of. How much lumber, wallboard, joint compound, whatever? Make a note of this information in your trusty notebook. Why? Because you cannot safely assume that all of the material delivered belongs to your job. A contractor may be doubling up on deliveries, planning to then deliver the overage to another job site and, quite ethically, bill you for yours and them for theirs. Not so ethically, some contractors in this situation have been known to bill the whole shipment to the first customer, while using some of it for another job and getting paid for that too.

Who Buys Materials— You or Your Contractor?

Traditionally, contractors have made a certain part of their profit by buying materials at a significant discount and passing some, but not all, of that saving on to the customer. This is a perfectly legitimate way to do business and need not be viewed with suspicion. Maybe some contractors would try to pull a fast one here, but you have done a scrupulous job checking your contractor out and devising a solid contract and you have little to fear.

If you are working with an architect or designer, he or she will of course be helping you to select and purchase many of the materials for the job, which is fine with the contractor, who has enough to do without going on shopping trips with you.

Even if you go it alone with no professional design help, you are totally within your rights to purchase for the project any appliances, fixtures, finishing materials, or any other item, and relive the contractor of that duty. But there are certain responsibilities and potential drawbacks that accompany your decision to do so.

First, you must be certain that all items supplied by you arrive at the site in plenty of time. You don't want to hold up the job for a few very expensive days and incur the wrath of the entire crew while everybody waits for the arrival of something that you ordered. Second, be aware that tracking down and ordering the items you want, and riding herd on the order to be sure it arrives on time, takes a lot of time and effort, probably more that you may really want to expend. Third, unless you stipulate differently in your contract, you will be responsible for the on-site storage and safety of everything you order and for returns, should they be necessary. Fourth, you should set aside a time when the contractor can look over the item you have ordered; he or she may notice a glitch that you might not catch.

It's a lot of work. Why do it unless you have some special access, not available to the contractor, to the items you want? It is rare that a homeowner can get a better price than a contractor, even in these days of discount home centers. And why not have your contractor order any special item? He or she may be able to get it at a better price.

More and more homeowners are buying their own materials today, which N'ann Harp thinks is about mistrust of contractors rather than any economic advantage. "When you do the math, you see that it makes no fiscal sense for homeowners to buy any materials hoping to save money. They rarely gain financially. But they may gain a measure of emotional protection from the fear that they're going to be taken advantage of."

LIEN RELEASES

No matter how much you trust your contractor, he or she must furnish you with lien releases from himself or herself, from the subcontractors who worked on the projects, and from the suppliers who delivered materials or equipment to the job site. These releases declare that the contractor paid for the services rendered and the supplies delivered and that you are not accountable for them. The documents release you from any further financial responsibility. You will obtain these releases at several points during the project, just before you make each scheduled payment to the contractor.

Here's why you must be concerned about this issue: Although it is unlikely that your contractor would gyp you, it is possible. You pay the contractor according to the payment schedule you have worked out. But instead of disbursing this money to the subs or suppliers, all of whom are waiting to be paid, a contractor could keep the money and become very scarce for a while or even leave town altogether with bills unpaid but a lot of cash in his jeans. Your careful research almost guarantees that the person you have hired is not unethical, but what about desperate? Has your contractor taken on so many jobs that he or she is stretched too thin and owes suppliers and subs from previous jobs? It may not be out-and-out theft but rather a question of robbing Peter to pay Paul. Even so, you don't want any part of it.

Lien releases from the contractor, subcontractors, and suppliers, stating that they have all been paid will give you peace of mind and an assurance that no one can come after you for money that you've already shelled out. A release of lien from the contractor alone is not sufficient. All that that says is that you have paid him or her; it says nothing about whether the others have been paid.

Remember, each time you make a scheduled payment to the contractor, get releases of lien from subs and suppliers. Your contractor may have release forms or you can accept a release written on the contractor's stationery letterhead or a plain sheet of paper, as long as it is properly signed.

Assuming that your contractor will pay the subcontractors and suppliers, you need to keep on top of this pesky lien business. To do so, use a manila folder to hold a list of suppliers and subcontractors who should be paid in each of the payment periods you and the g.c. have established.

BE THERE—EVERY DAY

You have the right to supervise and inspect the job. It's in your contract, or it should be. And to oversee the job effectively, you or a member of your family should look things over daily, armed with a project schedule, which will help you monitor progress. You'll also need a copy of the plans, a tape measure, and your trusty notebook. It's important that you do your inspections unobtrusively, say, at the end of the day when the crew has gone home or early in the morning before they arrive. You don't want to alienate your contractor by poking around suspiciously, officiously, and obsessively during working hours, getting in the way, asking unnecessary questions, and acting as though you're smarter than everyone else.

Most days, you won't need to say a word. You'll just be quietly pleased to see that things are going as they should. But one day you might notice that something's amiss. Perhaps the kitchen cabinets delivered that afternoon have the wrong finish—country pine, not the honey maple that you ordered. This is probably an easy fix. Just let the contractor know so that he can have them exchanged for the right ones. You may not even lose a day.

But if you hadn't been looking things over on a daily basis, half of the cabinets might have been installed before you noticed. What then? Sure, it was somebody else's fault. The supplier sent the wrong goods, and the contractor accepted them. So you would be within your rights to insist that the cabinets be removed, returned to the supplier, and replaced with the correct ones. It wouldn't cost you anything, but it could add at least a couple of days to the schedule, which in turn could throw other parts of the job off and cause more lateness down the line. Then you'd have a problem: Do you make do with the pine finish, which is all right but not, after all, what you ordered? Or do you exercise your rights, get the finish you specified, and set the job back? You wouldn't have this problem at all if you had been on the job and observant every day.

So what else do you look for on your solitary inspections? Check all appliances, lighting fixtures, flooring, and other materials to make sure their size, style, color are what you ordered and that there are no scratches or gouges. Then cast an eye over the workmanship. Is there sufficient insulation? Is it installed correctly? Are door and window openings in the right position? Are crucial measurements correct? Are electric outlets installed where you specified?

But what if you don't have the time, inclination, or skill for doing this kind of appraisal? Many of us would not be capable, even with the help of blueprints and other working drawings, to tell whether the toilet and whirlpool tub have been roughed in correctly or whether the framing is in perfect alignment. Reading about it may help. Libraries, bookstores, home centers, and hardware stores are prime sources for books that describe what happens in a typical remodeling or building project. There are many such books, geared for the nonprofessional yet detailed enough to help you conduct your inspections.

Hiring an on-site supervisor is another possibility. For a big job, some home-improvement experts strongly suggest taking this step, especially on a project that has no architect overseeing it. "It's a good idea," says attorney Ann Rankin, "to have someone to handle things in the field. Just be sure whoever you hire has no previous relationship with the contractor or any of the subs." This person, who may be an engineer, a contractor or architect serving here as a consultant, or another professional who has worked in the building industry, functions as a project administrator, checking the quality and building-code readiness of the work, assessing change orders, approving invoices for payment, and representing the owner in conflicts of interest between owner and contractor.

Although local building inspectors are assigned to do some of this work, Rankin says you cannot be sure how carefully they will do it. "Some inspectors are good; some are careless. But even if they overlook something very important, you cannot sue them—they have statutory immunity."

KEEPING THE JOB ON TRACK

The prospect is daunting. How can you, a nonprofessional with no experience in building or remodeling, stay on top of a complex project that may frighten you because it is so unfamiliar and is costing you so much money? You could always take the easy way out and turn everything over to the contractor, but, says Mike McClintock, who writes on home-improvement subjects for the *Washington Post,* "...that's too much like falling asleep in the barber's chair." McClintock thinks the best way to stay on track is to schedule weekly meetings to be attended by you, the general contractor, major subcontractors—such as plumbers, electricians, or cabinet installers—who are currently working on their part of the job, and any design professionals you may have hired.

When to Call Your Contractor

It's sort of like calling your doctor. Should you or shouldn't you? You don't want to be a pest, but you do want to protect your health and prevent future problems. Well, we can't help you with doctor-patient relationships, but here are some suggestions from the National Association of the Remodeling Industry (NARI) about when to pick up the phone and talk to your contractor.

- Electricity, plumbing, or other utilities are not working properly. You should also make a call if you are experiencing power surges.
- The job site poses hazards to homeowner, workers, or neighbors.
- A worker or subcontractor does not keep an appointment. Before you get annoyed, call the contractor and find out if there is a legitimate reason for the delay.
- The work is not progressing to your satisfaction. Maybe you hate the kitchen cabinets you chose from a catalog, or the bathroom turned out to be smaller than you expected. Whatever the problem, the sooner the contractor is notified the easier it will be to make changes.
- You or a member of your family is having a problem with one of the workers. A third-party perspective may be all that is needed to defuse the situation.

End-of-the-week meetings are best, says McClintock; and the idea is to uncover potential problems and delays by reviewing the previous week's work (including things that didn't get done) and discussing what is to be done in the week to come, including expected deliveries of supplies or materials.

Go into these meetings equipped with the project schedule, which will tell you if the job is where it should be. And to avoid delays—or traffic jams when too many subs and workers show up on the same day—review which sub is due next on the project, what that person is supposed to do, and how long it should take him or her to do it. This kind of discussion should go a long way toward circumventing lack of coordination, which McClintock calls the main stumbling block in a remodeling project. What the job needs, he says is "a logical sequence that prepares foundations for framers and framing for drywallers." If, for example, the electrician is missing before the rough wiring is finished, the contractor may not be able to schedule an inspection, which could throw the whole job off by creating a delay for drywall installers, finish carpenters, painters, and other tradespeople.

Delays, like death and taxes, are inevitable. And what makes delays all the more frustrating is that often they're not anybody's fault and you are thus denied the dubious comfort of blaming someone.

Schedules shift all the time, and it's best to shift with them. Plumbers, for example, are sometimes called away on emergencies. If this happens, the removal of your old bathroom sink and the installation of the new one will have to wait. Occasionally such things as a truckers' strike or an equipment breakdown prevent timely delivery of materials.

However, if delays are longer than a few days, or if they occur often, there is a problem. If it seems that subs are not on site when they should be, talk to your contractor about it. To circumvent delivery delays, you might also ask him or her to move the delivery schedule forward so that you are certain to have materials on hand.

At your end-of-the-week meetings you can also discuss changes in plan. Some of these will result from discoveries made during the renovation process and must be taken care of. Some will result from mistakes, such as somebody forgetting to order the windows, which must also be taken care of. "Any significant deviation from the blueprints needs a little last-minute planning because it wasn't scheduled initially," says McClintock.

INSPECTIONS

People from your local building department may be a frequent presence on-site; the bigger the project, the more often they'll be there. Inspections normally occur after important phases of work are completed—footings, foundation, framing, rough plumbing and wiring, insulation. All work that is scheduled to follow a part of the job that

Here's a Switch—Contractors' Advice to Homeowners

- Unless there is an emergency, direct all questions, instructions, and complaints to the contractor, not to subs, workers, or suppliers. But if the subs are about to cut a hole for a door where there should be a window, stop them right away. As soon as you've stopped them from incorrectly cutting that hole for a door, call the contractor and clear the matter up. Unless you follow this chain of command, you will create confusion, make people mad, and perhaps even contribute to a sloppy job. Similarly, pass on to the contractor any directives or complaints the subs address to you. This is simple professional courtesy and is strongly advised by all the contractors of our acquaintance.

  Retired contractor Stan Pritikin warns against trying to play subs and contractor against one another. "I've had customers tell a sub that I OK'd something that I never even heard about," says Pritikin. "For example, replacing one bathroom faucet with another that they decided they liked better—at no extra charge. Or installing a few additional outlets, also at no extra charge, of course. It never works because either I notice it myself or the sub tells me about it. Even better, the sub refuses to do it and sends the customer to me. The only thing that comes from this game-playing is a lack of trust, which is not a good way to do business."
- Talk about concerns and problems right away, even if you feel a little nervous about bringing them up. "If the owners and I can discuss a problem right away, it can almost always be taken care of," says veteran contractor Sven Swenson.
- Don't lose your temper with the contractor or subs. Of course you can't picture yourself indulging in such rude behavior; but after a few weeks of the stress and tension of a remodeling, you might be tempted to blow off steam when something goes wrong—and something will. Maybe one of the workers let your cat Fluffy out of the basement in spite of the agreement that the basement door would not be opened during working hours. Or perhaps there's a scratch in your expensive new porcelain tub because a worker dropped a tool into it. It's maddening, yes; but yelling, screaming, and insulting the contractor or workers (you'd be surprised how often this happens) will cast a gloomy pall over the job. This cloud of resentment and hurt feelings will take time to dispel even if you apologize, which you almost certainly will have to do.
- Have faith in the contractor you researched so carefully and hired with great confidence. If something puzzles you or looks wrong, by all means mention it, says Swenson. In the flurry of activity that happens during renovation, mistakes inevitably happen. The contractor, no matter how competent he or she is, may miss something. But, Swenson urges, do refrain from passing on advice you got from a neighbor or your brother-in-law or somebody on a home-improvement TV show about how a door should be hung or a cabinet installed.

requires an inspection has to wait until the inspection is conducted and the go-ahead given. However, if you have chosen your contractor wisely, he or she will take inspections into account and will arrange the job so that work continues somewhere on-site while you wait for the OK. If there are no problems, permission to proceed is usually granted immediately.

Although inspections sometimes delay a job, and inspectors can be high-handed and officious, they are necessary. A careful inspection may unearth a problem that could cause trouble, even danger, down the line to you and your family. However, if you observe, or hear from your contractor or other industry insiders, that the building department in your community is overwhelmed and inefficient, be cautious. This condition, quite common in major metropolitan areas, can produce overworked employees, long waits, and sloppy inspections. Try to be on-site whenever inspections are scheduled. If you notice an inspector—the one looking over the electrical work, for example—doing a fast, slipshod look-see or even signing the inspection card without actually looking at the work, talk to your contractor about your concerns. He or she should be knowledgeable enough about all aspects of the project to know whether the wiring has been done properly or not. Plus, your contractor has probably worked with the electrician many times and can vouch for the work.

If you're still concerned, you could bring an independent inspector on board, an electrician who is not associated with anyone working on your project, for example. Asking him or her to look over the wiring that the official inspector was so cavalier about may calm your fears, but it does have its downside. Unless your independent inspector can come in right away, the job will be held up and the entire work schedule could be thrown off. If you do decide to go this route, you could hire a retired local contractor or get recommendations for an independent inspector from a real estate or construction attorney.

PUNCH LIST

No, this is not a list of the reasons why you should smack your contractor in the mouth, and we hope you're not feeling like doing so at this point.

A punch list is an itemized list of details that remain to be done, usually small things such as installing hardware or touching up paint. It's drawn up by the customer toward the end of the job. Start making the list after your last walk-through with the contractor. Take your own walk-through and look carefully at the remodeled rooms in both natural and artificial light, making note of every detail that was part of the job but has not yet been done. You will probably have to negotiate some of these items when you review this list with your contractor. Some details may not be fixable, some may not be worth fixing, some don't belong on the list.

For instance, if a drawer on a newly installed kitchen cabinet sticks, that's a bona fide item. But if a cabinet drawer in a room that wasn't remodeled sticks, that's a different story. One of the workers may agree to fix it for you, but it's not a punch-list item and it'll cost you extra.

If you and your contractor don't see eye to eye on some of the items, punch-list time can be very tense. You want your house back; your contractor wants to get out of there and move on to the next job. Try to be as flexible here as you have been throughout the project. Negotiate what you can live with and what you can't.

Make a copy of the negotiated list, one for you, one for the contractor, and request that the items be taken care of as quickly as possible. You will get no argument from the contractor on this because final payment hinges on a completed list. The contractor will schedule a day or two when workers and some subs come back, do what they have to, and leave. Then, and only then, do you write the last check.

Here are some things to look for as you prepare your list:

- Do door windows, doors, and cabinets open easily and smoothly?
- Is all hardware in place?
- Are all appliances in working order?
- Does any of the new plumbing leak? Are hot and cold faucets working properly?
- Do new floors lay flat without bubbles or bulges?
- Is the grout even and neat on new tile floors?
- Does any of the paint need touching up?
- Is all of the molding installed securely and painted or stained? Are all exposed edges covered with molding?

FINAL PAYMENT

Everybody will be happy when you write the last check. But don't do it until everything is in place, including:

- A punch list completed to your satisfaction.
- A demonstration of the proper use of all your new appliances and other equipment, such as a water heater or an electrical service box.
- Manufacturer warranties on all the appliances and equipment that your contractor provided.
- Lien waivers from the contractor and all subcontractors.
- A cleanup of the construction site and the property per the specifications in your contract.
- The contractor's final invoice showing that the contract is paid in full.

The final payment—the amount of which you have already hammered out with your contractor and lawyer and have specified in the contract—should not be made the minute the paint dries. Some experts recommend delaying the last payment until 30 days (others say 60 days) after the job is completed, to allow time for problems to appear and be corrected.

And you've also got to wait for mechanics'-lien deadlines to expire. Any person filing a lien must notify the homeowner within a prescribed period of time, typically 30 to 60 days. Consumer advocates for the remodeling industry strongly suggest that customer hold back the last payment until that time period has elapsed. (For the lien time limits in your state, call the consumer-affairs division of your attorney general's office; see listings in Chapter 7.)

Withholding payment until you've got everything you're supposed to get sounds manipulative—and it is, really—but it is based on sound psychology. Most people will work better when they are still owed a significant portion of the money for which they contracted. If your contractor has been paid in full, what's the motivation for coming back promptly to take care of a leaking roof or a lopsided cabinet door? A sense of honor? A desire to protect a good reputation? Maybe. But you can't count on getting speedy results from a busy, maybe overbooked, contractor for reasons as philosophical as those.

You've got warranties on the work, of course, probably for a year. And he or she will almost certainly come back sometime. But what if you've got a problem now? How long do you want to live with it? Holding back some money—many experts recommend 15 to 20 percent; 10 percent if it's a big job—will get you the treatment you want.

A Summary of Important On-The-Job Do's and Don'ts —

- DO be friendly from Day One to Final Payment. If a dispute arises, try to see the issue from the contractor's point of view as well as your own. Your cooperative attitude will pay off even if a serious dispute arises.
- DO keep close track of all deliveries, making sure that the invoice matches what was delivered. Get a copy of the invoice and keep it in a safe place. But DON'T sign for any deliveries.
- DO know who each workers is and who is employing him or her. Be aware of what each person is working on and how much time they put in.
- DO ask the contractor to introduce you to subs and workers as they come on the job. Not only does this help to maintain a pleasant and friendly atmosphere, but it also makes it easier for you to know who is there on any given day and what they are working on. If a sub does work on your job and you are not aware of it, you will not be obtaining a mechanic's lien release from him or her. You could be in financial trouble if that sub is not paid by the contractor.
- DO consider hiring an engineer to inspect the installation of air conditioning, heating, plumbing, and other complex systems.
- DO get as much feedback as you can from your building-department inspectors as they look over the job. They are good sources of information on the quality of the work being done.
- DON'T pay for any materials that you have not received or for services that have not been rendered. Stick to the payment schedule spelled out in the contract, which is geared to completed work, not to dates.
- DON'T make any payments at all until you have lien releases from all subs and suppliers for work done to date.
- DO alert your neighbors to what's about to happen; make clear when the job will start, approximately when it will end, and whom they can contact if they have serious complaints.

- DO keep all kids—yours and any who come to visit—away from the Dumpster or piles of construction debris. This debris, which often contains sharp nails, shards of glass, toxic materials, and other dangers, is no place for kids to play, no matter how fascinating it seems to them.
- DO inspect—with the contractor—all cabinets, appliances, plumbing fixtures, and other material that arrives at the site. Significantly damaged goods will have to be sent back. If everything looks all right, draft and cosign a memo with your contractor that acknowledges their condition and store the items in a safe place. If, come installation time, you find damage, you will know it happened on the site.
- DO have on-site before the job starts any products (appliances, hardware, light fixtures, whatever) that you, not the contractor, are providing. Better to have the items there and in the way than to delay the whole job while everybody waits for them. (The wait for a major appliance could be six weeks or more. Besides, you can always find a place to store a few things, even if it's in a neighbor's garage.)

six

Troubleshooting

If you have followed the advice offered throughout this book, it is very unlikely that you will experience any serious trouble with the contractor you hired to oversee your remodeling project. Any problems that arise will probably be minimal, a far cry from the disasters often faced by people who approach this very serious endeavor carelessly and enter into it without protection.

Your early homework about the feasibility of the project and how much you can afford to spend on it, plus your careful checking of contractor candidates, your creation of a clear contract, your close supervision of the project, will add up to a successful and satisfying experience. Your efforts have already eliminated almost all possibility of major money problems, unethical or financially insolvent contractors, shoddy workmanship, uncooperative attitudes, and sloppy business practices.

That doesn't mean you won't have any problems. But it does mean that you have built an excellent foundation for dealing with the inevitable difficulties sanely and intelligently. One of the last things you want to do is to fire your contractor and hire another one to come in and finish the job, a so-called solution that will cost you a lot of time and money and extend the already difficult disruption of your home. The other last thing you want to do is end up in court.

FIRST, KEEP YOUR COOL

Having a remodeling blow like a hurricane through your home can be quite stressful. And to make matters a little more difficult, the stress tends to build as the job progresses. Demolition, the first step, is noisy and a bit scary, but it's impressive because you can see the dramatic changes as they happen. Foundation work and rough framing happen more slowly, and the changes are less dramatic. By now you may be feeling the stress of the noise, the mess, and the strangers wandering in and out. Maybe you start worrying about the money you're spending; maybe the crew seems to be having too much fun horsing around with one another or taking too many breaks. Finishing work, the last stage, takes the longest. This stage seems interminable and you're beginning to wonder if your life will ever go back to normal. If you're watching the money and invoices, as you should be, you are also being frequently reminded of how costly all of this finishing stuff is—ceramic tile, wood floors, Corian countertops, those handsome kitchen faucets you had your heart set on. You knew (because you did your homework) what all these items would cost; but now that it's really happening, it seems to upset you more. And the job is taking so long.

If your simmering anxiety and frustration bubble up into anger, trouble is not far behind. Losing your temper with the contractor or subs or laborers is bad enough. In a way you're at the mercy of these people, and you don't want them mad at you. Another danger of feeling grumpy and frustrated is that you may make a poor decision based on emotion rather than reason, such as insisting that a sub be fired or that the crew redo an important part of the job to rectify a minor mistake. Reactions like these can throw the project into jeopardy at a time when everyone's temper is already slightly frayed.

POTENTIAL PROBLEMS

Remodeling is not an exact science. It is almost impossible to go through an entire project without encountering some surprises, changes, delays, or cost overruns. Actually it is the inevitable surprises and changes that create the delays, and the delays in turn lead to the cost overruns. Don't worry; remember, you've built some flexibility into the project-schedule section of your contract to allow for these contingencies, so a few delays of a couple of days each will not throw the job into chaos.

If you are remodeling an older house—say, one 30 years old or older—expect the unexpected. Opening up a wall could reveal something unpleasant, such as faulty, old wiring or rotted framing members or another flaw that will have to be corrected before the job can proceed. Even in a newer house, you may come across surprises. Sven Swenson recalls a remodeling project he did in a house that was built in the 1970s. "We were adding circuits to the existing electrical service when we discovered that aluminum electrical wire had been used rather than copper. Copper was very expensive in the '70s when the house was put up," says Swenson, "and some people substituted aluminum, which was a lot cheaper. Trouble is, it caused a lot of fires and did terrible damage, and it's now outlawed, in fact. We didn't plan to rewire that house, but once we discovered the aluminum, we had to do it." And the homeowner had to pay for it.

Dealing with rotting posts discovered during the framing stage would not be a big problem; replacing them would take only day or two and would add only a few hundred dollars to the cost, if that. Rewiring the house is another matter; it's expensive and it could hold up the job for a couple of days or an entire week if the whole house needs rewiring. What do you do?

You've got to make the change because the old wiring is dangerous. And it would never pass an electrical inspection anyway. There's no one to blame for it; it couldn't have been predicted. Discuss it calmly with your contractor; this is where the good relationship you have built comes in. Can the electrician do the rewiring right away? How much extra will the electrician charge? Can work proceed on other parts of the project while the electrical disaster is being corrected? Is there a way to make up for the delay at other stages of the job? Is it possible to economize on other aspects of the job—substituting economical fiberglass bath fixtures for the pricey porcelain ones you originally ordered, for example—to offset the unexpected cost of the wiring? If you keep your cool and maintain your cooperative attitude, there's a way out of every problem.

Other changes in the project may be initiated by you (although your diligent preplanning will have definitely kept them to a minimum) or by a mistake that someone else has made. To keep the job on track and on budget, try to roll with the punches and restrain yourself from making any expensive and time-consuming alterations that may pop into your mind as you take your daily stroll through the project. One day, for example, looking at the holes that have been cut for the windows, you may say to yourself, "Why didn't I notice that these windows are a little small? Bigger ones would look so much better." Well, maybe they would, but indulging in this alteration will hold things up and cost you money, probably a couple of hundred dollars, maybe more. Is it really worth it? If you decide that it is, that the aesthetic pleasure you will get from larger windows far outweighs the consequences, go for it. But don't do this kind of thing too often—it will throw off the timing of the job, exasperate the entire crew, and threaten the harmonious relationship you have developed with your contractor. Some customers of Sven Swenson's made changes nearly every week. "After we installed the windows they initially wanted, they asked for larger ones," Swenson recalls. "They said they wanted a 'different experience' with the windows in that room. Well, they got a different experience, all right. We made all the alterations they asked for and charged them what the changes and delays cost, which made them mad, so they sued us. They lost."

You can also roll with the punches when a change must be made because of a mistake, such as accepting a finish for your kitchen cabinets that is different from the one you ordered, the example cited in Chapter 5. Correcting this mistake—which can be attributed to the supplier for sending the wrong product and the contractor for accepting it—would not cost you any money, but it would slow down the job. Removing the wrong cabinets and installing the right ones will take time; if you should have to wait for the right ones to be delivered from the manufacturer, it would cause an even longer delay. If you can let it go and live with the "wrong" cabinets, great. Everyone will appreciate your flexibility and the job will go forward. But don't do this unless you truly *can* live contentedly with your decision. You are entitled to get what you ordered, and if it's very important to you, you must insist. To minimize problems and bad feelings, do your insisting calmly and cooperatively. Resist all impulses to point the finger at someone in a hostile way.

Watch Out for the While-ya's

There's something about having a competent contractor and crew around the house that makes us notice all those little pesky problems we never found the time to fix. And in spite of your best efforts to plan your project and prepare for all eventualities, one day you may find yourself saying to the contractor, "While you're here could you just fix...?" Not so affectionately known as "while-ya's," these requests include such seemingly inconsequential tasks as repairing a busted window screen or installing a new porch light.

The tasks may be small or inconsequential, but their effect is not. They throw the job off schedule and off budget, and actually cost you more than they would have if you'd dealt with them separately. Don't throw the job out of whack for details such as these; instead, discuss them with the contractor, who may suggest that you hire one of the subs to come back for a day or two after the job is completed and take care of them.

Keep in mind that if it isn't in the contract, any added job, no matter how small, will raise the ante. Before you start a sentence with a while-ya, make sure the additional work is really worth it

Some while-ya's are actually good ideas, and it may make sense to incorporate them now, while the house is already torn up. But it doesn't make sense to plan them now. That should have been done long ago, in the early stages of your project.

Damage is another potential problem. Sometimes a member of the crew scratches, nicks, or otherwise damages a new item, such as your just-installed, shiny stainless-steel sink or an existing piece of furniture, say, your antique dining table. You're furious, and you want somebody to pay. But did you see it happen? Do you know for sure it was a crew member and not one of your kids or the dog? And unless you're keeping up with your daily inspections, you can't even be sure the sink didn't arrive that way. Of course, you'll want to talk to the contractor about the damage, but be reasonable. If he or she has done a good job for you so far, you don't want to be haggling over a minor repair. If, however, the damage is serious and it is certain that a worker caused it, then you must talk to the contractor about what he or she is willing to repair or replace.

According to many experts in the remodeling and construction fields, the most frequent and most bitter complaint about remodeling projects is that they don't finish on time. Delays, are inevitable; but if you stay on top of the project schedule and communicate regularly with your contractor, you may be able to keep them from getting out of hand.

Left alone, delays can escalate quickly and nastily. For example, say the new cabinets for your kitchen will be a few days late, so the cabinet installer has decided to go to another job, promising to return in a day or so. Then you discover that the bathroom tiles are in that same late shipment, which means that the bathroom-tile installer can't go to work yet either. The cabinets arrive, but the installer now has a slipped disk and doesn't know when he'll be back. At this point, things are so far out of whack that the schedule for other important subcontractors has been disrupted and they have had to go to other jobs too, to return who knows when? The job is still limping along, but it's chaotic and inefficient and the work will soon suffer.

You can head off this nightmare scenario by dealing with the delays as they happen and taking steps to correct them before the job swerves out of control. Instruct the contractor to call another cabinet supplier, for example, or insist that the contractor hires a substitute cabinet installer to replace the one who is sick.

EASY SOLUTIONS, OR HOW TO STAY OUT OF COURT

Who wins when you go to court? That's right, the lawyers. The litigation process is costly, time-consuming, upsetting, and worst of all, not always satisfying, even if you win a judgment. For one thing, a contractor who is told by the court to return to the job and finish it doesn't have much incentive to do his or her best work. And if you are awarded a monetary settlement, it may take a long time to collect; in fact, if the contractor skips town, you'll likely never see a penny. Even if you are able to collect, you will probably find that the settlement doesn't amount to anywhere near enough to hire a new contractor and finish the project in good order.

So when things start going wrong on the job, seek a solution rather than a lawsuit. Begin working toward the solution by talking to your contractor as soon as you notice a problem or feel uneasy about something. Communicate in as friendly and evenhanded a way as possible, letting the contractor know that you view him or her as part of the solution, not part of the problem. Explore ways that you can get the job back on track or solve your dispute together, without outside help. Be flexible and open to compromises, even if doing so will cost you a little money. Listen to what the contractor has to say; work at being receptive to his or her point of view.

If things are particularly tense at this point, you may be tempted to shut down the job and pay off or out-and-out fire the contractor. Resist these temptations for the moment. After careful deliberation and mediation, you may decide to take one of those actions at a later date; but for now, you should be working toward a peaceful resolution.

If, in spite of your best efforts to be calm, cooperative, and fair, you and your contractor find yourselves unable to settle a dispute effectively, let your next step be toward an amicable alternative dispute resolution, not toward an adversarial day in court. That is why most responsible remodeling contracts include arbitration clauses. Yours, if you followed our advice, includes provisions for both mediation *and* arbitration. But even if your contract includes no such clauses, all is not lost. You would probably be surprised to learn how many organizations and agencies in your community offer third-party negotiation for all kinds of disputes, including remodeling problems. Call your local Better Business Bureau, your state attorney general's consumer-protection office, the local bar association, or the American Arbitration Association.

Dispute resolution is an appropriate step for any homeowner-contractor conflict, large or small, that cannot be resolved amicably.

Mediation, generally the first step taken toward resolution, is a gentle process, which is typically run like a round-table discussion. A neutral third party encourages and facilitates the resolution of the dispute between homeowner and contractor. Mediation is typically a voluntary undertaking, and although it seems somewhat contradictory to insist on it in a legal document, it can still be very effective if both parties agree to its inclusion in the contract. Another reason for the effectiveness of mediation is that it can be scheduled very quickly so that you can start the process as soon as you come up against a particularly sticky conflict, while you and your contractor are still speaking to each other and before your disagreement has had time to fester and grow.

In this informal, nonadversarial process, a trained facilitator helps the parties to identify the issue clearly, to try joint problem-solving, and to come up with creative solutions. The mediator may offer suggestions and point out issues that he or she thinks have been overlooked, but resolution of the dispute rests with the parties themselves. Although compromise is usually required, a skillfully handled mediation often leaves both parties feeling like winners. And, says a spokesperson for United States Arbitration and Mediation, "Because in the end they have created the solution themselves, the parties feel more satisfied with this than with a settlement that is imposed on them by a court, even if they're not totally thrilled with the results."

And mediation works. According to figures released by the American Arbitration Association, 85 percent of commercial matters and 95 percent of personal injury matters end up in written settlement agreements as a result of mediation.

But, if for some reason mediation does not do the trick for you and your contractor, your contract also contains a provision for binding *arbitration*. And if your contract does not for some reason include this provision, you can, and should, seek arbitration anyway.

Arbitration is a slightly more formal process than mediation, more like a hearing than a round-table discussion. There's no judge or jury, but both parties present their cases to a carefully selected neutral arbitrator who will listen, ask questions, weigh the issues, and make a decision. Although arbitration takes place outside the courts, it nonetheless results in a final and legally authoritative decision that is similar to a court judgment. And because your contract provision calls for *binding* arbitration, you must accept and abide by the decision made by the arbitrator, which can be enforced by the courts if necessary. Many preprinted remodeling contracts specify that the American Arbitration Association handle the process, but there are other organization that do this work as well.

You're Fired!

Firing your contractor is an extreme measure, and the only legal reason for doing so is equally extreme: an unjustifiable refusal by the contractor to perform according to the terms of the contract, typically by refusing to perform, failing to perform, or performing late or barely. All of this unpleasant behavior is otherwise known as breach of contract. (By the way, your contractor can fire you under the same breach of contract provision, which on the part of the homeowner usually means failure to pay as stipulated in the contract.)

To stand up in court, your charges of refusal to perform must be verified against the contract, which is just one more good reason to keep careful records as the job proceeds and to document any changes to the contract with both parties' signatures. Although this extreme measure will cause you some problems—a possible battle in court, the expense and hassle of hiring a new contractor to finish the job—it is right and necessary if the contractor you've got consistently and unjustifiably refuses to perform the work or is unable to perform it competently.

We feel sure, however, that you will never have to face such grim realities as firing your contractor or going to court. Following the advice we have outlined in this book will protect you from these and the other disasters and horror stories you have heard about, most of them caused, as you now know, by a few bad-apple contractors and, more often, by homeowners who did not take the time to be careful. Sure, you'll have a couple of problems along the way. But taking your time with the selection process, insisting on a solid contract, and staying on top of the job will produce all the conditions you need for a successful and satisfying remodeling.

seven

Resources

State Attorneys General 162
Better Business Bureaus 166
State Contractor Regulatory Offices 171
Arbitration Services 176
Building Codes 177
Contracts 178
Federal Trade Commission (FTC) 179
Web Sites 180
Home Inspections 181
Insurance 182
Financing 183
Booklets and Video 184
Professional Remodeling Organizations 185

Before you hire a contractor, check with the following agencies and organizations, listed by state, for information that will help you make an intelligent decision. Have any complaints been filed against any of your candidates, have these complaints have resolved, how long has the company been in business, do your candidates possess the licenses that are required in your state? These organizations also offer helpful consumer-protection information.

STATE ATTORNEYS GENERAL

ALABAMA
11 South Union Street
Montgomery, AL 36130
334-242-7300

ALASKA
Post Office Box 110300
Juneau, AK 99811-0300
907-465-3600

ARIZONA
1275 West Washington Street
Phoenix, AZ 85007
603-543-4266

ARKANSAS
200 Tower Building
323 Center Street
Little Rock, AR 72201-2610
501-682-2007

CALIFORNIA
1300 I Street
Suite 1740
Sacramento, CA 95814
916-324-5437

COLORADO
1525 Sherman Street
Denver, CO 80203
303-866-3052

CONNECTICUT
55 Elm Street
Hartford, CT 06141-0120
860-808-5324

DELAWARE
820 North French Street
Wilmington, DE 19801
302-577-8400

DISTRICT OF COLUMBIA
441 Fourth Street NW
Washington, DC 20001
202-727-6248

FLORIDA
The Capital, PL 01
Tallahassee, FL 32399-1050
850-487-1963

GEORGIA
40 Capital Square SW
Atlanta, GA 30334-1300
404-656-4585

HAWAII
425 Queen Street
Honolulu, HI 96813
808-586-1282

IDAHO
State House
Boise, ID 83720-1000
208-334-2400

ILLINOIS
100 West Randolph Street
Chicago, IL 60601
312-814-2503

INDIANA
402 West Washington Street
Indianapolis, IN 46204
317-233-4386

IOWA
1305 East Walnut
Des Moines, IA 50319
515-281-3053

KANSAS
120 SW 10th Avenue, 2nd Floor
Topeka, KS 66612-1597
785-296-2215

KENTUCKY
State Capitol, Room 116
Frankfort, KY 40601
502-696-5300

LOUISIANA
Post Office Box 94095
Baton Rouge, LA 70804-4095
225-342-7013

MAINE
State House Station Six
Augusta, ME 04333
207-626-8800

MARYLAND
200 St. Paul Place
Baltimore, MD 21202-2202
410-576-6300

MASSACHUSETTS
1 Ashburton Place
Boston, MA 02108-1698
617-727-2200

MICHIGAN
Post Office Box 30212
525 West Ottawa Street
Lansing, MI 48909-0212
517-373-1110

MINNESOTA
State Capitol, Suite 102
St. Paul, MN 55155
612-296-6196

MISSISSIPPI
Post Office Box 220
Jackson, MS 39205-0220
601-359-3692

MISSOURI
207 West High Street
Jefferson City, MO 65101
573-751-3321

MONTANA
215 North Sanders
Helena, MT 59620-1401
406-444-2026

NEBRASKA
Post Office Box 98920
Lincoln, NE 68509-8920
402-471-2682

NEVADA
100 North Carson Street
Carson City, NV 89701
775-687-4170

NEW HAMPSHIRE
25 Capital Street
Concord, NH 03301-6397
603-271-3658

NEW JERSEY
25 Market Street
Trenton, NJ 08625
609-292-4925

NEW MEXICO
Post Office Drawer 1508
Santa Fe, NM 87504-1508
505-827-6000

NEW YORK
Department of Law, The Capitol
Albany, NY 12224
518-474-7330

NORTH CAROLINA
Post Office Box 629
Raleigh, NC 27602-0629
919-716-6400

NORTH DAKOTA
600 East Boulevard Avenue
Bismarck, ND 58505-0040
701-328-2210

OHIO
30 East Broad Street
Columbus, OH 43226-0410
614-466-3376

OKLAHOMA
2300 North Lincoln Boulevard
Oklahoma City, OK 73105
405-521-3921

OREGON
1162 Court Street NE
Salem, OR 97301
503-378-4320

PENNSYLVANIA
Strawberry Square
Harrisburg, PA 17120
717-787-3391

RHODE ISLAND
150 South Main Street
Providence, RI 02903
401-274-4400

SOUTH CAROLINA
Post Office Box 11549
Columbia, SC 29211-1549
803-734-3970

SOUTH DAKOTA
500 East Capitol
Pierre, SD 57501-5070
605-773-3215

TENNESSEE
500 Charlotte Avenue
Nashville, TN 37243
615-741-1671

TEXAS
Post Office Box 12548
Austin, TX 78711-2548
512-463-2191

UTAH
State Capitol, Room 236
Salt Lake City, UT 84114-0810
801-538-1326

VERMONT
109 State Street
Montpelier, VT 05609-1001
802-828-3171

VIRGINIA
900 East Main Street
Richmond, VA 23219
804-786-2071

WASHINGTON
P.O. Box 40100
1125 Washington Street SE
Olympia, WA 98504-0100
360-753-6200

WEST VIRGINIA
1900 Kanewha Boulevard East
Charleston, WV 25305
304-558-2021

WISCONSIN
Post Office Box 7857
Madison, WI 53707-7857
608-266-1221

WYOMING
State Capitol
Cheyenne, WY 82002
307-777-7841

BETTER BUSINESS BUREAUS

There are nearly 200 offices of the Better Business Bureau in the United States, too many to list here. This abbreviated list provides one per state. Call that Bureau—or the Council of Better Business Bureaus at 703-276-0100—for the office nearest you.

ALABAMA
Montgomery
312 Montgomery Street
4th Floor
Montgomery, AL 36104-3559
334-262-5606

ALASKA
Fairbanks
542 Fourth Avenue
Suites 217 & 219
Fairbanks, AK 99707
907-451-0222

ARIZONA
Phoenix
4428 North 12th Street
Phoenix, AZ 85104-4585
602-240-3973
($3.80/call charged to caller's credit card)
24-hour line 900-225-5222 (95¢/min.)

ARKANSAS
Little Rock
1415 South University
Little Rock, AR 72204-2605
501-664-7274
800-482-8448 (Arkansas only)

CALIFORNIA
Sacramento
400 S Street
Sacramento, CA 95814-6997
916-443-6843

COLORADO
Denver
1780 South Bellaire, Suite 700
Denver, CO 80222-4350
24-hour line 303-758-2100

CONNECTICUT
Wallingford
821 North Main Street Ext.
Wallingford, CT 06492-2420
24-hour line 203-269-2700

DELAWARE

Wilmington

1010 Concord Avenue
Wilmington, DE 19802
302-594-9200

DISTRICT OF COLUMBIA

1012 14th Street NW
Washington, DC 20005-3410
24-hour line 202-393-8000

FLORIDA

Jacksonville

7820 Arlington Expressway
Suite 147
Jacksonville, FL 32211
904-721-2288

GEORGIA

Atlanta

Post Office Box 2707
Atlanta, GA 30301
404-688-4910

HAWAII

Honolulu

1132 Bishop Street, 15th Floor
Honolulu, HI 96814-3801
24-hour line 808-536-6956

IDAHO

Boise

1333 West Jefferson
Boise, ID 83702-5320
208-342-4649

ILLINOIS

Chicago

330 North Wabash Avenue
Chicago, IL 60611
312-832-0500
($3.80/call charged to caller's credit card)
24-hour line 900-225-5222
(95¢/min., 24 hr.)

INDIANA

Indianapolis

22 East Washington Street
Suite 200
Indianapolis, IN 46204-3584
317-488-2222

IOWA

Des Moines

505 Fifth Avenue
Suite 950
Des Moines, IA 50309-8137
515-243-8137

KANSAS

Topeka

501 SE Jefferson
Suite 24
Topeka, KS 66607-1190
785-232-0454

KENTUCKY

Louisville

844 South Fourth Street
Louisville, KY 40203-2186
24-hour line 502-583-6546
24-hour line 800-388-2222
(KY and southern IN only)

LOUISIANA

Baton Rouge

2055 Wooddale Boulevard
Baton Rouge, LA 70806-1546
504-926-3010

MAINE

Portland

812 Stevens Avenue
Portland, ME 04103-2648
207-878-2715

MARYLAND

Baltimore

2100 Huntingdon Avenue
Baltimore, MD 21211-3215
24-hour line 900-225-5222
(95¢/min.)

MASSACHUSETTS

Boston

20 Park Plaza
Suite 820
Boston, MA 02116-4344
617-426-9000
800-422-2811
(802 area code only)

MICHIGAN

Grand Rapids

40 Pearl NW, Suite 354
Grand Rapids, MI 46503-3001
616-774-8236; 24-hour line
800-684-3222
(Western MI only)

MINNESOTA

Minneapolis-St. Paul

2706 Gannon Road
Minneapolis-St. Paul, MN
55116-2600
24-hour line 651-699-1111
24-hour complaint line
800-646-6222

MISSISSIPPI

Jackson

4500 I-55 North
Jackson, MS 39211
601-987-8282

MISSOURI

St. Louis

12 Sunnen Drive, Suite 121
St. Louis, MO 63143
24-hour line 314-645-3300

NEBRASKA

Lincoln

3633 O Street, Suite 1
Lincoln, NE 68510-1670
24-hour line 402-476-8855

NEVADA

Las Vegas

Post Office Box 44108
Las Vegas, NV 89116-2108
702-320-4500

NEW HAMPSHIRE

Concord

410 South Main Street
Concord, NH 03301-3483
603-224-1991
603-228-3789, 3844

NEW JERSEY

Trenton

1700 Whitehorse–Hamilton Square Road
Trenton, NJ 08690-3596
609-588-0808

NEW MEXICO

Albuquerque

2625 Pennsylvania NE
Albuquerque, NM 87110-3657
505-346-0110
800-873-2224 (NM only)

NEW YORK

New York

257 Park Avenue South
New York, NY 10010
212-533-6200 ($3.80/call charged to caller's credit card)
24-hour line 900-225-5222 (95¢/min.)

NORTH CAROLINA

Raleigh

3125 Poplarwood Court
Raleigh, NC 27604-1080
919-872-9240
800-222-0950 (eastern NC only)

OHIO

Columbus

1335 Dublin, Suite 30A
Columbus, OH 43215-1000
614-486-6336

OKLAHOMA

Oklahoma City

17 South Dewey
Oklahoma City, OK 73102-2400
405-239-6081

OREGON

Portland

333 SW Fifth Avenue
Portland, OR 97204
24-hour line 503-226-3981
800-488-4155 (OR and southwestern WA only)

PENNSYLVANIA

Pittsburgh

300 Sixth Avenue
Pittsburgh, PA 15222-2511
24-hour line 412-456-2700

RHODE ISLAND

Warwick

120 Lavan Street
Warwick, RI 02888-1071 (Providence)
401-785-1212 (inquiries)
401-785-1213 (complaints)

SOUTH CAROLINA

Columbia

2330 Devine Street
Columbus, SC 29205
803-254-2525

TENNESSEE

Nashville

201 Fourth Avenue North
Nashville, TN 37219-1778
24-hour line 615-250-4222

TEXAS

Austin

2101 South IH 35
Suite 302
Austin, TX 78741-3854
24-hour line 512-445-2911

UTAH

Salt Lake City

1588 South Main Street
Salt Lake City, UT 84115-5382
24-hour line 801-487-4656
800-456-3907 (UT only)

VERMONT

Massachusetts

800-422-2811 (from 802 area code only)

VIRGINIA

Richmond

701 East Franklin
Suite 712
Richmond, VA 23219-2332
24-hour line 804-648-0016

WASHINGTON

Spokane

508 West Sixth Avenue
Spokane, WA 99207-2356
509-455-4200

WEST VIRGINIA

Charleston

Post Office Box 2541
Charleston, WV 25329
304-345-7502

WISCONSIN

Milwaukee

740 North Plankinton Avenue
Milwaukee, WI 53203-2478
414-273-1600 (inquiries)
414-273-0123 (complaints)

STATE CONTRACTOR REGULATORY OFFICES

ALABAMA
Home Builder's Licensure Board
400 South Union Street
Montgomery, AL 36130-3605\
334-242-2230

ALASKA
Division of Occupational Licensing
Post Office Box 110806
Juneau, AK 99811-0806
907-465-2546; 907-465-3035

ARIZONA
Registrars of Contractors
800 West Washington
Phoenix, AZ 85007
602-542-1525; 888-271-9286 (toll-free in Arizona)

ARKANSAS
Contractors Licensing Board
621 East Capitol Avenue
Little Rock, AR 72202
501-372-4661

CALIFORNIA
Contractors State License Board
Post Office Box 26000
Sacramento, CA 95826
916-255-3900; 800-321-2752 (automated system)

COLORADO
Colorado Electrical & Plumbing Board
1580 Logan Street
Denver, CO 80203-1941
303-894-2300

CONNECTICUT
Department of Consumer Protection
Licensing Services
165 Capitol Avenue
Hartford, CT 06106
860-566-2825

DELAWARE
Division of Revenue
Department of Finance
Post Office Box 2340
Wilmington, DE 19899
302-577-5800 ext. 7522

DISTRICT OF COLUMBIA
Department of Consumer and Regulatory Affairs
614 H Street NW
Washington, DC 20001
202-727-7090

FLORIDA

Construction Industry
Licensing Board
7960 Arlington Expressway
Jacksonville, FL 32211-7467
904-727-3689

GEORGIA

Construction Industry
Licensing Board
166 Pryor Street SW
Atlanta, GA 30303
404-656-2448

HAWAII

Contractor's Licensing Board
Post Office Box 3469
Honolulu, HI 96301
808-586-2700

IDAHO

Public Works Contractors State
Licensing Board
Post Office Box 83720
Boise, ID 83720-0073
208-327-7326

ILLINOIS

Department of Professional
Regulation
100 West Randolph
Chicago, IL 60601
312-814-4500

INDIANA

Division of Consumer Services
402 West Washington Street
Indianapolis, IN 46204-2746
317-232-6201

IOWA

Workforce Development,
Div. of Labor
1000 East Grand Avenue
Des Moines, IA 50319-0209
515-281-3606

KANSAS

Department of Revenue
Division of Taxation Registration
915 SW Harrison Street
Topeka, KS 66625-0001
785-296-3081

KENTUCKY

Division of Occupations
and Professions
Department of Housing &
Developing
Capitol Annex, Box 456
Frankfort, KY 40601
502-564-8644

LOUISIANA

Licensing Board for Contractors
Post Office Box 14419
Baton Rouge, LA 70898-4419
504-765-2301; 800-256-1392

MAINE

Office of Licensing
and Registration
35 State House Station
Augusta, ME 04333-0059
207-624-8500

MARYLAND

Department of Licensing and Registration
500 North Calvert Street
Baltimore, MD 21202-3651

MASSACHUSETTS

State Board of Building Regulations and Standards
1 Ashburton Place, Room 1301
Boston, MA 02108
617-727-3200

MICHIGAN

Bureau of Construction Codes
Post Office Box 30245
Lansing, MI 48909
517-241-9254

MINNESOTA

Department of Commerce
Licensing Division
133 East Seventh Street
St. Paul, MN 55101
612-296-6319

MISSISSIPPI

Board of Contractors
2001 Airport Road
Jackson, MS 39208
601 354 6161

MISSOURI

Division of Professional Registration
Post Office Box 1335
Jefferson City, MO 65102-1335
573-751-0293

MONTANA

Department of Labor & Industry
Post Office Box 8011
Helena, MT 59604-8011
406-444-7734; 800-556-6694

NEBRASKA

Department of Labor
13134 Farnam Street
Omaha, NE 68102-1898
402-595-3095; 402-595-3189

NEVADA

State Contractors Board
4220 South Maryland Parkway, Bldg. D
Las Vegas, NV 89119
702-486-1100

NEW HAMPSHIRE

Consumer Protection & Antitrust Bureau
Office of the Attorney General
33 Capitol Street
Concord, NH 03301-6397
603-271-3641

NEW JERSEY

Bureau of Homeowner Protection
CN 805
Trenton, NJ 08625-0805
609-530-8800; 609-530-8801

NEW MEXICO
Department of Regulation and Licensing
Post Office Box 25101
Santa Fe, NM 87501-5101
505-827-7030

NEW YORK
Department of State
Division of Licensing Services
41 State Street
Albany, NY 12231
518-473-2492

NORTH CAROLINA
Licensing Board for General Contractors
Post Office Box 17187
Raleigh, NC 27619-7187
919-571-4183

NORTH DAKOTA
Contactors Licensing Division
Office of the Secretary of State
600 East Boulevard Avenue
Bismarck, ND 58505-0500
701-328-3665

OHIO
Construction Industry Examining Board
6606 Tussing Road
Post Office Box 4009
Reynoldsburg, OH 43068
614-644-3493

OKLAHOMA
Occupational Licensing Division
State Department of Health
1000 NE 10th Street
Oklahoma City, OK 73117
405-271-5217

OREGON
Construction Contractors Board
700 Summer Street, NE
Salem, OR 97309-5052
503-378-4621 ext. 4900

PENNSYLVANIA
Bureau of Professional & Occupational Affairs
Post Office Box 2649
Harrisburg, PA 17105-2649
717-787-8503

RHODE ISLAND
Contractor's Registration Board
Building Code Commission
1 Capitol Hill
Providence, RI 02908-5859
401-222-1270

SOUTH CAROLINA
Residential Builders Commission
Post Office Box 11329
Columbia, SC 29211-1329
803-734-4255

SOUTH DAKOTA
Department of Revenue
Business Tax Division
445 East Capitol Avenue
Pierre, SD 57501-3185
605-773-3311

TENNESSEE

Department of Commerce and Insurance
500 James Robertson Parkway
Nashville, TN 37243-1150
615-741-8307; 800-544-7693

TEXAS

Department of Licensing & Regulation
Post Office Box 12157
Austin, TX 78711
512-463-2906

UTAH

Division of Occupational & Professional Licensing
Post Office Box 45805
Salt Lake City, UT 84145-0805
801-530-6628

VERMONT

Department of Labor & Industry
Drawer 20, National Life Building
Montpelier, VT 05620-3401
802-828-2107

VIRGINIA

Department of Professional & Occupational Regulation
3600 West Broad Street
Richmond, VA 23230-4917
804-367-8511; 804-367-2945 – Tradesman Office

WASHINGTON

Department of Labor & Industries
Post Office Box 44450
Olympia, WA 98504-4450
360-902-5202

WEST VIRGINIA

Contractor Licensing of Labor
319 Building #3
Capitol Complex
Charleston, WV 25305
304-558-7890

WISCONSIN

Department of Commerce
Post Office Box 7082
Madison, WI 53707-7082
608-261-8500

WYOMING

Division of Labor Standards
6101 Yellowstone
Cheyenne, WY 82002
307-777-7261

ARBITRATION SERVICES

American Arbitration Association
335 Madison Avenue
New York, NY 10017
800-778-7879
www.adr.org

American Homeowners Foundation (AHF)
6776 Little Falls Road
Arlington, VA 22213-1213

For information on the RESOLVED dispute resolution program, call 800-489-7776

Council of Better Business Bureaus
4200 Wilson Boulevard
Arlington, VA 22203
703-276-0100

For information about BBB dispute resolution, contact your local bureau. If there is no bureau near you, call the Council or visit the web site www.bbb.org

United States Arbitration & Mediation
600 University Street
Suite 2000
Seattle, WA 98101
800-318-2700
800-933-6348
www.usam.com

More information on dispute resolution is available from your state bar association or attorney general's office.

BUILDING CODES

For information about building codes and how to satisfy them, contact any of the following organizations:

Building Officials & Code Administrators International (BOCA)
4051 West Flossmoor Road
Country Club Hills, IL 60477
708-799-2300

International Code Council
5203 Leesburg Pike
Falls Church, VA 22041
703-931-4533
www.intlcode.org

International Conference of Building Officials (ICBO)
5360 South Workman Mill Road
Whittier, CA 90601
888-699-0541
www.icbo.org

CONTRACTS

The following organizations offer information about, and sample contracts for, residential construction and remodeling projects:

American Homeowners Foundation (AHF)
6776 Little Falls Road
Arlington, VA 22213-1213
703-536-7776; 800-489-7667
www.americanhomeowners.org

American Institute of Architects (AIA)
1735 New York Avenue NW
Washington, DC 20006
For contacts and other documents, call 800-365-2724

Associated General Contractors of America
333 John Carlyle Street
Alexandria, VA 22314
703-548-3118
www.AGC.org

Smart Consumer Services
2111 Jefferson Davis Highway
Suite 722 North
Crystal City, VA 22202
703-416-0257
www.SConsumer.com

FEDERAL TRADE COMMISSION (FTC)

The Federal Trade Commission (FTC) helps consumers find the appropriate agencies for help and counsel on many consumer-related matters, including remodeling contractors, how to hire them, and how to work with them.

Call: 877-FTC-HELP for the regional office nearest you.

Write: Consumer Response Center
Federal Trade Commission
600 Pennsylvania Avenue NW
Washington, DC 20580

Visit: www.ftc.gov

WEB SITES

Here's a smattering of web sites that deliver general advice and information on remodeling, do-it-yourself projects, and related matters. Some also claim to have lists of prequalified contractors. Well, maybe. Make a note of the names of the contractors near you if you like. But be sure to check every name carefully as we suggest elsewhere in this book.

http://architecture.about.com
http://homerepair.about.com
www.contractors.com
www.contractorfind.com
www.cslb.ca.gov
www.findhomebuilding.com
www.housenet.com
www.homefix.com
www.improvenet.com
www.nari.com
www.nationalcontractors.com
www.pueblo.gsa.com
www.remodel.com
www.remodelingresource.com
www.remodeling.hw.net
www.RemodelingCorner.org
www.repair-home.com
www.Sconsumer.com
www.ThatHomeSIte.com
www.thecontractorpages.com
www.ThisOldHouse.com

HOME INSPECTIONS

American Society of Home Inspectors (ASHI)
932 Lee Street
Des Plaines, IL 60016
847-290-1919; 800-743-ASHI
www.ashi.com

National Academy of Building Inspection Engineers (NABIE)
Box 520
611 York Street
York Harbor, ME 03911
207-351-1915; 800-294-7729
www.nabie.org

INSURANCE

These organizations can provide information on the insurance you need before—and after—you undertake a remodeling.

Insurance Information Institute
Publication Service Center
110 William Street
New York, NY 10038
800-942-4242
www.iii.org

Insure.com
76 La Salle Street
West Hartford, CT 06107
860-233-2800
www.insure.com

FINANCING

These resources can help you find the facts and figures you'll need to make a decision about financing your project.

American Bankers Association
1120 Connecticut Avenue NW
Washington, DC 20036
800-BANKERS
www.aba.com

DFS Mortgage Service
16501 Sherman Way
Van Nuys, CA 91406
800-334-7004
www.remodelingcash.com

Federal Housing Administration (FHA)
A division of Housing and Urban Development (HUD)
451 Seventh Street SW
Washington, DC 20410
800-733-4663

Federal National Mortgage Association (Fannie Mae)
Public Information Office
3900 Wisconsin Avenue NW
Washington, DC 20016
800-7-FANNIE

The web boasts a mind-boggling array of information on home-equity loans. Some sites provide excellent information and worksheets; others are strictly advertisements, so be careful. Here are some worthwhile sites.

www.financenter.com
www.hsh.com/heqsample.html
www.bankrate.com
www.moneyminded.com

BOOKLETS AND VIDEOS

How to Resolve Consumer Disputes
Road to Resolution: Handling Consumer Disputes
Home-Equity Credit Lines
Warranties and *Service Contracts*

All four are free from the:
Federal Trade Commission Public Reference
Washington, DC 20580
202-326-2222

When Your Home Is on the Line: What You Should Know About Home-Equity Loans

Available free from:
Publications Services
Mail Stop 127
Federal Reserve Board
21st and C Streets
Washington, DC 20551
Fax: 202-728-5886

You Can Build It! a 58-page booklet about the permit process is available for $4.50 from the International Conference of Building Officials. Call 888-699-0541 for your copy.

Invest in a Dream, a video prepared by the American Institute of Architects, is available from the AIA for $21.95. Call 800-365-2724 to order a copy.

PROFESSIONAL REMODELING ORGANIZATIONS

These trade organizations can direct you to their member remodeling professionals in your area.

American Association of Architects (AIA)
1735 New York Avenue NW
Washington, DC 20006
202-626-7300
www.e-architect.com

American Institute of Building Design
991 Post Road East
Westport, CT 06880
800-366-2423
www.aibd.org

American Society of Interior Designers (ASID)
608 Massachusetts Avenue NE
Washington, DC 20002-6006
202-546-3480
www.interiors.org

National Association of the Remodeling Industry (NARI)
4900 Seminary Road
Alexandria, VA 22311
703-575-1100
www.nari.org

National Kitchen & Bath Association (NKBA)
687 Willow Grove Street
Hackettstown, NJ 07840
908-852-0030
www.nkba.org

Remodelers Council of the National Association of Home Builders
1201 15th Street NW
Washington, DC 20005
800-868-5242
www.remodelingresource.com

Glossary

Your contractor, subcontractors, and others involved with your project will be sprinkling their conversation with words that may not be familiar to you. Here are a few of the words you're likely to hear. Learning what they mean can help bring you up to speed.

Abandonment. The failure of both parties to abide by the terms of a contract they have entered into.

Acoustical board or tiles. Material used on ceilings and walls to deaden or control sound.

Allowances. Projected costs for specific parts of a job or for the materials used in completing those parts. When a substitute material is used, the portion of its cost that is more, or less, than the originally specified material is added to or subtracted from the total price. Also, a sum of money set aside in the construction contract for items that have not been selected and specified. Allowances are best reserved for items that will not impact earlier stages of construction—choosing ceramic tile for a floor, for example, may require a change in framing or underlayment material.

Amperes (amps). Standard unit for measuring the rate at which electrical current flows. Circuit breakers are rated to handle a specific amperage, typically 200-amp service in new construction.

Appraisal. An inspection that estimates the value of property on the current real estate market.

Arbitration. The settlement of a dispute with the help of a mutually selected impartial party. In contractor-related disputes, the arbitration usually binds both parties to accept the findings of the arbitrator.

Asbestos. A fibrous building material that was once widely used because of its stability and resistance to fire. It has since been discovered that asbestos, now outlawed, is associated with lung disease.

Asphalt. A bituminous material used as a waterproofing element in shingles.

Assessment. An official valuation of property for the purposes of taxation.

Awning window. A sash that is hinged at the top and opens out.

Backfill. Earth, or earth mixed with gravel, used to fill the excavation around a foundation; also, the act of filling in around a foundation.

Backhoe. An excavating machine with a bucket for scooping out earth.

Backsplash. A protective panel—often of ceramic tile—installed to protect the wall behind a counter or sink from stains and water marks.

Balloon construction. A type of house construction in which the studs are continuous from foundation to roof.

Balloon payment. Due at the end of a fixed period of time, a balloon payment pays off the entire principal of a debt. It is typically preceded by a series of installment payments that pay off the accruing interest on the debt.

Balusters. Vertical supporting members of a stair rail, usually decorative and placed close together.

Balustrade. The entire stair rail, including handrail, balusters, and posts.

Baseboard. A finish board used to cover the joint at the intersection of wall and floor.

Batts. Blankets of fiberglass insulation usually sandwiched between heavy paper or foil and generally sized to fit between framing members in a wall.

Beadboard. Wood paneling, often used as wainscoting, that features strips of wood separated by a narrow projecting molding called a bead.

Beam. A straight structural member that resists transverse loads. Also, an inclusive term for joists, girders, and rafters.

Bearing wall. A wall that supports part of the structure; also called a structural, or load-bearing wall.

Bid. A document stating the amount of money for which a contractor will perform a construction, remodeling, or repair job.

Blind-nail. A method of fastening that conceals the nails.

Blown-in insulation. Loose insulation that is inserted with a blower into attics, crawl spaces, or walls.

Blueprints. Working plans, prepared by an architect or designer and printed in blue ink on a white ground, or white ink on a blue ground, that show all aspects of the construction to be done in a building or remodeling project.

Board. Lumber 8 inches or more in width and less than 2 inches thick.

Board and batten. Vertical siding with battens (narrow strips of wood) covering the joints of the boards.

Board foot. A unit of lumber that measures 1 foot square and 1 inch thick.

Bond. A guarantee that the bond insurer will pay for losses incurred by a homeowner as a result of a contractor's failure to perform or to meet contract requirements. See also completion bond (page 190); performance bond (page 198).

Brad. A thin, short finishing nail.

Bridging. Diagonal bracing added between floor joists for strength.

Builder grade. A term that indicates a product—such as windows, doors, cabinets—of average quality and reasonable price.

Building codes. National, state, and local requirements that regulate building materials, construction techniques, and building occupancy in the interest of safety and public health.

Building paper. Thick, water-repellent paper used to insulate and dampproof a building before the siding is installed.

Building permit. Official permission issued by the local government allowing construction or renovation to begin.

Bullnose. A trim tile with a convex radius on the outer edge, used for finishing kitchen or bath countertops, topping off wainscoting, or turning an outside corner.

Cabinets—custom. Cabinets that are tailor-made to suit the specifications of a particular customer.

Cabinets—stock. Cabinets that are built in quantity and in a limited range of sizes and shapes.

Cantilever. Any part of a structure that projects beyond its main support and is balanced on it.

Casement window. With the sash hinged at the vertical edge, a casement window opens outward by means of a mechanical crank.

Casework. Assembled cabinetry or millwork.

Casing. Trim around a door, window, or other opening.

Caulking. A waterproof, putty-like material used to seal cracks, crevices, and seams.

Certificate of occupancy. A document that permits the legal habitation of a dwelling. CO's are issued by the local code-enforcement authority when all building-code requirements have been met. Without this important document, a homeowner would be unable to sell the house.

Chair rail. Molding that runs horizontally along a wall about 3 feet above the floor (or chair height); often used to the cap the upper edge of wainscoting. Originally designed to prevent chair backs from damaging a wall, chair rails are now largely decorative.

Change order. A written agreement that amends a remodeling contract and includes such information as changes in materials, prices, and cost of the work to be done. A change order should be as detailed as the original contract and should be signed by both parties.

Circuit breaker. A switchlike device located in the electrical service panel or circuit-breaker box that shuts down power to parts, or all, of the house and limits the amount of power flowing through a circuit; the up-to-date equivalent of a fuse.

Clapboard. Siding made up of horizontal boards tapered so that the thicker side is exposed and the thinner side is lapped under the next course of boards.

Clear. Term used in wood-grading systems to describe the best grade. See also common (page 190); select (page 200).

Closed grain. Wood with small and loosely spaced pores.

Code. If something is built "to code," it satisfies the requirements of the building code, which regulates the practices and materials of the construction industry.

Collateral. Something of value that is used to secure a loan. The collateral is then held as security until the loan is paid off.

Common. Term used in wood-grading systems to describe a material that contains visual flaws that do not impair structural integrity. See also clear (page 190); select (page 200).

Completion bond. If a remodeling contact is not fulfilled, a completion bond guarantees that the insurer will pay the additional funds needed to complete the job as specified by the contract. See also performance bond (page 198).

Comprehensive service. The most complete type of service offered by an architect in a home-improvement project. It includes all aspects, from preliminary design consultation to overseeing the daily progress of the project.

Concrete. A building material used for foundations; concrete is made of portland cement, sand, gravel (or crushed stone), and water.

Concrete board. A panel made of mortar and fiberglass, often used as a backing for ceramic tile. Also known as Wonderboard.

Contractor. A person with whom you sign a contract to build you a house or an addition, renovate your existing house, or perform any other sort of home-improvement work for an agreed-upon sum of money. A contractor may do the work single-handedly or may contract with others (subcontractors) and manage their work. See also specialty contractor (page 201); subcontractor (page 201).

Corner board. A decorative board used to trim and reinforce exterior corners of walls.

Cornice. The exterior detail at the meeting of an exterior wall and roof overhang; the molded projection that finishes the top of a building. Inside the house, a cornice is a decorative piece of wood or other material that covers the topmost part of a window, concealing curtain hardware.

Course. A horizontal layer of masonry units, one unit high; a horizontal layer of shingles.

Cove molding. A concave-shaped molding.

Crawl space. In a house without a standard basement, a shallow space between the first-story floor joists and the ground.

Cripple stud. A short stud used to frame spaces above and below window or door openings.

Crown molding. Decorative trim used to ease the transition between wall and ceiling.

Deck. The form on which concrete for a slab is placed; the floor or roof slab itself. The surface, installed over the supporting framing members, to which the roofing is applied. Also, a roofless outdoor-living area, constructed of wood, that adjoins a house.

Decking. The material used to build an interior floor system or an exterior deck.

Demo work. Demolition; the process of dismantling or destroying existing conditions.

Dimension lumber. The most common type of lumber for framing; dimension lumber for framing is milled to standard sizes of 3/4 to 1 1/2 inches thick and 2 1/2 to 11 1/4 inches wide.

Distribution box. A metal, wall-hung box that contains circuit breakers and connections to the service wires, and delivers electric current to all the outlets in a building.

Dormer. A structure that projects out from a sloping roof to form another roofed area that increases headroom and admits light through one or more windows.

Double glazing. Two parallel sheets of glass with an air space between them.

Double-hung window. A window in which two sashes slide vertically by each other.

Downspout. A pipe that allows water to drain from roof gutters; also called a leader.

Draw. A partial and scheduled payment that a contractor receives over the duration of a project as stipulated in the contract.

Drip cap. A piece of molding placed over a door or window to direct rainwater away from the opening.

Drip edge. A metal strip placed on roof edges to provide support for the overhang of the roofing material.

Dry rot. Powdery residue of wood that results from fungus destruction due to excess moisture.

Drywall. Panels with a paper covering over a gypsum core, that are used to cover interior walls and ceilings. (The panels are nailed or screwed onto the framing members, and the joints are taped and covered with joint compound.) Also called gypsum board, plasterboard, and Sheetrock.

Dumpster. A large metal container used on construction sites to hold rubbish.

Easement. A right or privilege that allows one person to use for a specific purpose property belonging to another.

Eaves. The lower edge of a roof that projects over the walls.

Elevation. A drawing that illustrates each side of a building or its interior wall in a straight-on view; also the sides of the building themselves, as in the "rear elevation" or the "eastern elevation."

Equity. The value of a property above the total amount of debt—such as a mortgage, lien, or home-equity loan—that is still owed on it.

Estimate. A prediction of the cost of performing work; an approximation of construction costs.

Exposure. The surface area of a board, shingle, roofing tile, or other building element that will be seen after installation.

Facade. The exterior face of a building.

Face. The better-looking side of a piece of wood or the side that is exposed when installed.

Face-nail. To nail at a right angle to a surface, with the nail heads exposed.

Fascia. Flat, horizontal boards that form a band around the edge of a roof.

Fiberboard. A prefabricated sheet of building material made of compressed wood or plant fibers; most commonly used as an underlayment for interior walls.

Final completion. The absolute completion of a contracted job; the point at which there is no more work to be done and final payment is appropriate. See also substantial completion (page 201).

Finish carpentry. That part of the carpentry process that pertains to interior elements such as doors, windows, moldings, and other decorative trim.

Finishing nail. A thin nail with a small head designed for setting below the surface of the finish material.

Firsts and seconds. The best grade of hardwood lumber.

Fitting. A faucet or any other mechanism that controls the amount of water entering and leaving a fixture.

Fixture. Any of the various parts of the plumbing or electrical systems that are installed permanently in a building, such as bathtubs, sinks, toilets, and wall or ceiling lighting fixtures.

Flashing. Any material used at intersections such as roof valleys, dormers, and above windows and doors to prevent the entrance of water.

Floor joists. Support beams, commonly installed parallel to other beams, to create a floor system to which a subfloor is attached.

Floor plan. Architectural drawing that shows the location of the rooms in a building, drawn from the point of view of an observer looking down at the structure.

Footing. A foundation for a wall, column, or chimney made wider than the object it supports.

Footprint. The perimeter and inside area of the foundation.

Foundation. That portion of a building or a wall that supports the superstructure.

Frame construction. A method of building in which the structural parts are wood or depend on a wood frame for support.

Framing. The rough, wood and/or metal structural skeleton of a building, including interior and exterior walls, floors, ceilings, and roof; also the structural members around a window or door opening.

Framing lumber. Used exclusively for framing, this rough lumber has been sawed, planed, and cut to length without further manufacturing or refining.

Frostline. The depth to which frost penetrates into the ground in a particular area; footings must be placed below this depth.

Furring strip. Strips of lumber spaced at desired intervals for the attachment of wall or ceiling covering.

Gable roof. A roof that has two sloping sides meeting in a ridge. The triangular section of a wall at each end of a gable roof is called the gable end.

Gambrel roof. A roof with two pitches—steeper on the lower slope and flatter toward the ridge—on each side of center; designed to provide extra room on upper floors.

General contractor. A person who enters a contract with a homeowner to oversee a construction or remodeling project and to coordinate the work of subcontractors. A general contractor is usually skilled in rough carpentry and knowledgeable about all areas of construction and may perform some of the construction work on the project. See also specialty contractor (page 201); subcontractor (page 201); tradespeople (page 202).

Girder. A heavy beam used to support wall beams or floor joists.

Glass block. A hollow masonry unit made of glass; used in residential construction to provide privacy without blocking the flow of light.

Glazing. The act of fitting with glass; also, the glass itself.

Glue-laminated lumber (Glulam). Large beams made by gluing smaller-dimension pieces of lumber together side-to-side.

Ground fault interrupter (GFI). A special electrical outlet, typically used near a water source, that protects against electric shock by instantly detecting a short circuit and automatically shutting off the power.

Grout. A thin mortar used to fill small joints or cavities, such as the spaces between ceramic tiles.

Gutter. The trough that channels rain water from a roof to the downspouts.

Gypsum board. A wall sheathing product made by encasing gypsum in a heavy paper wrapping; also called drywall, plasterboard, and Sheetrock.

Half-round. A molding with the shape of a half circle.

Handsplit shingle. A shingle made by splitting a block of wood, usually cedar or redwood, along its grain and thereby creating a shingle for roofing or siding; also called a shake.

Header. A structural member that spans the top of a window, doorway, or other opening in a wall.

Hip roof. A roof that slopes up to the center from all sides, requiring a hip rafter at each corner.

Hopper window. A sash that is hinged at the bottom and swings in and down.

Housewrap. A thin, paper air barrier that covers the sheathing of a house and is permeable to water vapor.

HVAC. Designation for the heating, ventilation, and air conditioning systems of a building.

I-beam. Prefabricated-metal structural member that resembles a capital *I* when viewed head-on.

Inspection. Before a certificate of occupancy can be issued, an official of the local building authority must inspect the completed remodeling project to ascertain that it meets local requirements.

Isometric. A drawing that shows three surfaces of an object in one view.

Jalousie window. A window made of movable horizontal slats of glass.

Jamb. Sides and tops of a window or door frame.

Joint compound. Also referred to as drywall mud, this puttylike material is spread on Sheetrocked interior walls to cover seams, nail heads, and screw heads, then later sanded to provide a smooth, even surface for painting.

Joist. A light framing member that supports the floor or the ceiling; joists run perpendicular to beams.

Kick plate. A metal plate or strip that runs along the bottom edge of a door to protect the finish.

Kiln-dried. Term used to describe lumber that has been dried in huge ovens called kilns.

Knee wall. Short wall under a low eave.

Lally column. A steel post, often adjustable, that is used to support a beam or girder.

Lath. A wood strip, metal mesh, or gypsum board that acts as a backing or a reinforcing agent for a plaster scratch coat or initial mortar coat. An application of lath and plaster was the most common way to finish a wall prior to the introduction of drywall.

Liability insurance. A type of insurance that protects a homeowner from lawsuits as a result of injuries or losses suffered by others on the homeowner's property. To protect themselves against these claims—or claims brought by a contractor's workers—homeowners should insist that any general contractor working for them carry liability insurance.

License. A certificate issued by state or local government that permits a contractor to work in that locality. Licensing may require only a fee to the state, or it may entail passing an examination, presenting proof of competence, or showing proof of insurance.

Lien. In the home-improvement field, a lien is a legal claim placed on real or personal property in lieu of payment for a supplier's debt (materialman's lien) or a worker's debt (mechanic's lien). Until liens are paid off, the property in question cannot be sold; in some instances, the lien holder can force the sale of the property in order to pay off the debt.

Lien waiver. A document signed by suppliers and workers that acknowledges payment and releases their right to place a lien against the property in question.

Light. An individual pane of glass or an opening for a pane of glass. A window with six panes in a sash is called a six-light window. See also muntin (page 197).

Limited service. An arrangement by which an architect or interior designer is hired to handle only selected parts of a project, such as design consultation or drawings.

Linear feet. A unit of measure that determines the distance between two points in straight line.

Lintel. Horizontal support member over an opening; also called a header.

Louver. Horizontal slats installed in a window or other opening; louvers are angled to exclude light, rain, and wind but permit the passage of air. Also, a kind of slatted window.

Low-E. Abbreviation for low-emissivity, a characteristic of window glass that has been treated with a reflective metallic coating to reduce the transfer of radiant heat from one side to the other.

Lumber. Wood cut from logs and then milled to form boards and planks suitable for use as structural members in construction.

Mansard roof. A roof with two pitches on each side, with the lower slopes steeper than the upper.

Masonry. Any construction of stone, brick, tile, concrete, or a similar material.

Mechanical systems. The plumbing, heating, and cooling equipment in building, which also includes ventilating fans and water heaters.

Mechanic's lien. See lien (page 196).

Mediation. A voluntary process in which a neutral third party helps two disputing parties to find a mutually acceptable solution. Usually attempted before more stringent solutions such as arbitration or litigation.

Membrane. A sheet material that is impervious to water or water vapor.

Millwork. Wood products—such as moldings, doors, windows, stairs—that have been manufactured for the interior of a building.

Molding. An ornamental strip of wood, plaster, or plastic used for finishing and decoration, and placed where interior surfaces meet, such as between walls and ceilings and around door and window openings.

Mullion. The vertical support member that separates a row of windows or panels in a door.

Muntin. Slender strips of wood that divide glass into individual panes, or lights. Some contemporary single-pane windows are available with a snap-in grid that creates the look of muntins.

Newel post. The upright post that supports the handrail of a set of stairs.

Nonconforming. A new house or an improvement to an existing house that is dissimilar to surrounding properties in age, size, style, or use. The term is sometimes used to indicate that a house, or a proposed renovation of a house, does not conform to current zoning ordinances.

Ogee. A molding with an S-shaped curve.

On center (OC). Term used to describe the distance from the center of one framing member to the center of the next.

Overhang. Portion of a roof that projects beyond the wall.

Panel. Large sheet of building material.

Particleboard. A building panel composed of small particles of wood and resins bonded together under pressure. Not suitable for use on the exterior or any place that comes into contact with water, particleboard is often used as an inexpensive alternative to plywood for subflooring or sheathing or as a base material for plastic-laminate kitchen counters.

Performance bond. Secured by the general contractor, this type of bond guarantees that the contract will be performed. See also completion bond (page 190).

Piers. Masonry columns used to support the structure of a house.

Pilaster. A column built within, but projecting slightly from, a wall for reinforcement and decoration.

Pitch. Term used to describe the steepness of a roof.

Plank. Lumber that is 6 inches or more wide and from 1 1/2 to 6 inches thick.

Plasterboard. A building panel consisting of a gypsum core enclosed in heavy paper; also referred to as drywall, gypsum board, and Sheetrock.

Plate. In a frame building, the top horizontal piece of the wall upon which the roof rests.

Platform framing. In this construction method, the subfloor provides a surface upon which walls for the next story are erected.

Plumb. Exactly perpendicular to the floor.

Plywood. A frequently used building material, plywood is made of thin sheets of wood glued together with the grain of adjacent layers at right angles to each other. It comes in 4- by 8-foot panels in a variety of thicknesses and in grades from rough to fine, and is available in both interior and exterior types. Plywood can be used for sheathing, subflooring, cabinets, paneling, and more.

Points. The charge levied by a lender for a loan, above and beyond the interest charges, points being the percentage points charged. Thus for a $100,000 loan, a loan fee of 1.5 points (that is, 1.5 percent) would require the borrower to pay $1,500 to the bank for the loan itself.

Post. A timber set on end to support a wall, girder, or other member of the structure.

Post-and-beam construction. A method of house construction in which the weight of the structure is supported by the perimeter walls and not by any interior walls.

Pressure-treated lumber. Ordinary wood, such as pine, that has been treated with chemicals under pressure to yield a product that is resistant to decay, fire, or insects; often used for the parts of a building that will come in contact with soil and water.

Punch list. Written inventory of items still to be completed near the end of a construction or remodeling project.

PVC pipe. Made of polyvinyl chloride, these pipes are used as waste lines in most new plumbing installations; they will not deteriorate and are considerably less expensive than copper, which was once widely used.

Quarter-round. A common form of molding that looks like one-quarter of a cylinder when viewed in cross section.

Rafter. A sloping framing member that supports a roof; the rafters make up the main body of the roof's framework.

Rebar. Abbreviation for reinforcing bar, a metal rod used to reinforce concrete.

Remodeling contractor. A general contractor who specializes in remodeling or renovation projects.

Ridge. The intersection of two sloped-roof planes; the apex.

Ridge board. The horizontal framing member that defines the ridge; the part of the roof frame into which the upper ends of the rafters are fastened.

Rigid insulation. Thermal insulating material formed to a flat board shape.

Rise. The height of a roof measured vertically from the base to the ridge.

Riser. A vertical member in a stairway, covering the space between stair treads.

Roof sheathing. The first layer of a roof, which is fastened to the rafters and used to support the final roofing material.

Rough carpentry. The preliminary framing, boxing, and sheathing of a wood-frame structure.

Rough-in. The stage at which plumbing, electricity, and equipment for heating and air-conditioning systems are installed; the installation of material before the walls are closed in.

Rough opening. A wall opening in a wall in which windows or doors will be installed.

R-value. The effectiveness or resistance of an insulating or building material wood, glass, masonry, insulation. A high R-value indicates high resistance to heat flow; for example, the walls in a new house typically hold R-19 insulation; the ceilings, R-28.

Sash. The movable part of a window; the frame in which panes of glass are set. A double-hung window has two sashes, both of which are operable.

Scaffold. An elevated temporary work platform; also called *staging*.

Schematic designs. Preliminary drawings and cost estimates provided by an architect at the earliest stage of project design.

Scratch coat. The first of three coats in the application of plaster.

Seasoned lumber. Construction lumber that has been dried to reach a suitable moisture content.

Section. Drawing that shows a vertical cut through an object or part of an object.

Select. The second-best grade of wood. See also clear (page 190); common (page 190).

Septic tank. Holding tank for a private sewage system, which is installed with a leaching field. Solid material settles to the bottom of the tank while liquid matter drains into the leaching field.

Service panel. The metal box into which the main electrical service cable is connected and from which the wiring of a house is routed through circuit breakers or fuses.

Setback. Minimum distance between a structure and property lines, as specified by local regulations. To build closer to the property line, a homeowner must obtain a variance from the zoning board, usually dependent on proving good cause.

Shake. A type of wood shingle. See handsplit shingle (page 194).

Sheathing. The rough covering applied over the framing on the outside of the house but underneath the final siding or roofing material. Also called sheeting.

Shed roof. A roof that has a single sloping side.

Sheetrock. Trademark for a building panel with a gypsum core inside paper layers; also called drywall, gypsum board, or plasterboard.

Shell. The framework or exterior structure of a building; a building with an unfinished interior. The shell is the support structure for all other materials.

Shiplap siding. Boards, applied horizontally, with edges that have been rabbeted to overlap flush from one board to the next.

Siding. The finish layer that covers exterior walls.

Sill plate. The horizontal framing member that is placed directly atop the foundation (also called a mudsill); the lowest structural member of a door or window frame.

Skirting. Any material, often plywood, that covers openings at the bottom of a structure.

Slab, or slab on grade. Concrete floor about 4 inches thick that is placed directly on the ground or on a gravel base.

Sleeper. A strip of wood laid over a concrete floor to which finish flooring is fastened.

Soffit. The underside trim portion of a cornice or any other overhang.

Specialty contractor. A remodeling or building contractor who works in one specialized trade such as plumbing, electricity, roofing, siding, masonry, or painting.

Specifications. The written or printed construction details for a remodeling or new construction project, describing exactly what will be done and what material will be used.

Square. The amount of material that will cover 100 square feet of roof area.

Stud. Vertical member used to frame walls; studs are placed at regular intervals to provide support and a nailing surface for wall coverings and exterior siding.

Subcontractor. Independent tradesperson hired for a specific part of a building or remodeling project such as roofing, plumbing, or painting.

Subfloor. A wood floor, typically plywood or particleboard, that is laid over the floor joists and over which the finished floor is installed.

Subpanel. Secondary electrical service panel installed to serve a specific zone of a house or an addition to the house.

Substantial completion. The point at which a remodeling is complete enough to allow occupancy although minor details remain to be done.

Supplier. The person or firm that provides materials for a building, home improvement, or repair project.

Taping. The process of preparing drywall for finishing. The tape, applied with joint compound, hides nail or screw heads and the seams where the sheets of drywall meet; the compound is then sanded until smooth to provide a suitable finish for painting.

Timber. Large pieces of lumber over 5 inches in thickness and width.

Time and materials (T&M). A method of billing a remodeling or construction job that stipulates paying the contractor a specific amount for the time to do a job, plus covering the contractor's actual cost for materials. Sometimes referred to as cost plus.

Toenail. Nail driven diagonally into the end of a framing member.

Tongue and groove. Lumber that has been machine cut so that the jutting edge of one board (tongue) fits into the grooved edge of another board (groove) for a nearly seamless fit.

Tradespeople. Also referred to as specialty contractors, tradespeople are carpenters, electricians, plumbers, and other trained construction professionals.

Trimming out. The stage of construction at which the final trim elements are installed.

Truss. An assembly of wood or wood-and-metal members, often in the shape of a triangle, used to support roofs or floors.

Underlayment. Material placed on a subfloor to provide a smooth, even surface for the application of resilient flooring. Also, a waterproof layer placed between the roof deck and the finished roofing material.

Vapor barrier. A material that blocks drafts and water; applied to sheathing before siding is put on and under concrete floor slabs that are on grade.

Variance. Granted by a local zoning board, a variance is an exception to a zoning ordinance.

Volts. Standard unit for measuring the quantity of force in an electrical current; 110 volts (for normal household lighting) and 220 volts (for major appliances) are the standard voltages for household currents.

Waferboard. A manufactured wood panel made of wood chips and glue; often used as a substitute for plywood in exterior-wall and roof sheathing.

Wainscoting. A facing, usually of wood or ceramic tile, that is applied only part way up a wall and capped by a chair rail; the upper part of the wall is usually painted or wallpapered.

Wall sheathing. Sheets of a building material, typically plywood, nailed to the studs as a base for exterior siding.

Workers' compensation insurance. Protection for workers who are injured on a job while performing their duties. To avoid a lawsuit over a worker's injury, homeowners who are engaged in home-improvement projects should insist that their general contractors have such coverage for all members of the crew.

Working drawings. Documents that give enough information for tradespeople to work from.